Prerna Bansal
Maths in Chemistry

Also of Interest

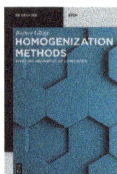

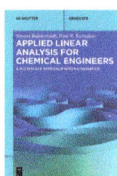

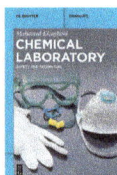

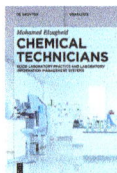

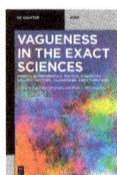

Prerna Bansal

Maths in Chemistry

Numerical Methods for Physical and Analytical Chemistry

2nd, Completely Revised and Extended Edition

DE GRUYTER

Author
Dr. Prerna Bansal
University of Delhi
Rajdhani College
Mahatma Gandhi Marg
110015 Delhi
bsprerna@yahoo.co.in

ISBN 978-3-11-133392-2
e-ISBN (PDF) 978-3-11-133444-8
e-ISBN (EPUB) 978-3-11-133450-9

Library of Congress Control Number: 2024935940

Bibliographic information published by the Deutsche Nationalbibliothek
The Deutsche Nationalbibliothek lists this publication in the Deutsche Nationalbibliografie;
detailed bibliographic data are available on the Internet at http://dnb.dnb.de.

© 2024 Walter de Gruyter GmbH, Berlin/Boston
Cover image: Who_I_am/iStock/Getty Images Plus
Typesetting: Integra Software Services Pvt. Ltd.

www.degruyter.com

Dedicated
To
My Parents

Acknowledgement

I owe this book to all those who encouraged me in some way or another to write this book. The source of inspiration for writing this book came from my students. I entirely dedicate this book to my beloved Parents who love me immensely and unconditionally and without whom I would have not been able to start writing a book in the first place. They have been a source of constant strength, support and love in difficult times. I am highly indebted for their constant patience for bearing with me while I was writing this book.

I would also like to express my heartfelt gratitude to my brother Mayank for putting up with me when I used to make odd asks. I also owe a special thanks to my two lovelies Divit and Ishaani for their beautiful smiles and laughters which used to melt my stress in a blink. This acknowledgement would have been incomplete without mentioning my beautiful sister Sonia for her excellent advice at all times when when she was busy with her Ph.D work. I express my gratitude to all my students who believed in me and kept asking me to write a book which can provide all the mentioned topics under a single wing. I would also like to express my sincere gratitude to all my teachers who taught me and inspired me to be the one. I also express a sincere word of thanks to my Husband Anup Goel who kept boosting my morale during the final stages of this manuscript completion for when I met him first. I really owe him for his constant support and encouragement during the entire process of its proof reading. I also owe a word of thanks to my in laws during the proof reading of this book.

I express my sincere thanks to the publisher De Gruyter's for their cooperation and processing of this manuscript. I also want to express a special word of thanks to Emma
Svensson, Ute Skambraks and Dipti Dange for their constant help and support. Above all, I would like to say thanks to Thakurji for providing me courage, motivation, strength and above all, his grace for giving me enough courage for writing this book.

During the revision of second edition of this book, I owe a great deal of gratitude to my lovely daughter Miraya who is my lifeline. Her smile and laughter has been a source of constant love and encouragement to me. Apart from my parents, I would like to dedicate this second edition of this book to my daughter Miraya also.

I would also like to express my sincere thanks to Ishwarya Mathavan for her patience and support during the revision of this book.

June 2024

https://doi.org/10.1515/9783111334448-202

Contents

Chapter 1
Units and measurements

1.1 Introduction

In science, especially physical chemistry, quite a lot of measurements are involved. A physical quantity holds a great importance. A physical quantity is the product of two quantities, that is, a unit and a number. While measuring any parameter in physical chemistry, number itself is not relevant unless it is hooked to a unit. For instance, 6 metre has the number "6" and unit "metre". Also, while correlating two physical quantities, constant and variables are involved. A constant is a physical quantity that has a fixed value, for example, Planck's constant ($h = 6.626 \times 10^{-34}$ J s) and Avogadro's constant ($N_A = 6.023 \times 10^{23}$ mol^{-1}). A variable, on the other hand, is a quantity that can have any permissible values based on some other quantities for which the given variable is a function, for example, p is a function of V, T, n, that is,

$$p = nRT/V \tag{1.1}$$

where p can be represented as $p = f(V, T, n)$.

Further, the variables are classified as dependent and independent variables. In the earlier example, since p is a function of V, T and n, p is a dependent variable while the rest are independent variables.

1.2 Basic arithmetic operations

A physical quantity is often the result of correlation of other physical quantities that are related by mathematical operators. To do so, basic arithmetic operations are performed on these quantities. We are all familiar with basic mathematical operations: addition, subtraction, multiplication and division. In this section, we will simply sum up the basics learned so far (Table 1.1).

Table 1.1: Mathematical operations.

Operation	Expression	Ordering
Addition	$A + B$	$A + B = B + A$
Subtraction	$A - B$	$A - B \neq B - A$
Multiplication	$A \times B$ or $A.B$ or AB	$A \times B = B \times A$
Division	$A \div B$ or $\dfrac{A}{B}$	$\dfrac{A}{B} \neq \dfrac{B}{A}$

https://doi.org/10.1515/9783111334448-001

Some of the equations used in physical chemistry involving these operations are listed below:

$$\text{Ideal gas equation: } pV = nRT \tag{1.2}$$

$$\text{Henderson–Hasselbalch equation: } \mathrm{pH} = \mathrm{p}K_a + \log(\mathrm{acid}/\mathrm{salt}) \tag{1.3}$$

$$\text{Einstein equation: } E = mc^2 \tag{1.4}$$

$$\text{Nernst equation: } E = E^0 - (RT/nF)\log Q \tag{1.5}$$

$$\text{Onsager equation: } \wedge = \wedge^0 - b\sqrt{c} \tag{1.6}$$

$$\text{Arrhenius equation: } k = A\exp(-E_a/RT) \tag{1.7}$$

1.3 Hierarchy of operations

Before performing arithmetic operations, there are some set of rules that should be followed. In conventional mathematics, BODMAS rule is used. They are also called operator precedence or rules of precedence or order of operations. A general method of remembering them is used in different places. For example, "PEMDAS" is the acronym used in the United States. The order of operations is as follows:
(1) Parentheses
(2) Exponential
(3) Multiplication
(4) Division
(5) Addition
(6) Subtraction

According to these rules, parentheses precede exponents which precede multiplication and division. While solving parentheses, the functions inside the parentheses are solved inside the parentheses itself, and is then expanded further. Also, the other operators are solved from left to right. For example,

a. $2 \times 3 + 4 = 6 + 4 = 10$
But $2 \times (3 + 4) = 2 \times 7 = 14$

b. $5 + 4^{1/2} = 5 + 2 = 7$
But $(5 + 4)^{1/2} = 3$

c. $(4 \times 8) \div 2 = 32 \div 2 = 16$. But $(4 \times 6 \div 2) = 24 \div 2 = 12$

Index rule

According to the index rule, index denotes the times a number has to be multiplied by itself, where index is a number which is raised to a power. The rule can be illustrated below.

(a) $a^x \times a^y = a^{x+y}$ e.g. $2^3 \times 2^5 = 2^8$

(b) $a^x / a^y = a^{x-y}$ e.g. $2^5 / 2^3 = 2^2$

(c) $(a^x)^y = (a^y)^x = a^{xy}$ e.g. $(2^3)^2 = 64$

(d) $a^{-x} = 1/a^x$ e.g. $(2)^{-1} = 1/2 = 0.5$

(e) $(ab)^x = a^x \times b^x$ e.g. $(2 \times 3)^4 = 2^4 \times 3^4 = 1296$

(f) $\left(\frac{a}{b}\right)^x = \frac{a^x}{b^x}$ e.g. $\left(\frac{2}{3}\right)^4 = \frac{2^4}{3^4}$

(g) $(a+b)^n \neq a^n + b^n$ e.g. $(2+3)^4 = 625$ while $2^4 + 3^4 = 97$. Hence, both are not equal.

1.4 Classification of unit system

The history of unit systems begins with the British Imperial System of units, followed by US Customary and International System of Units (metric system). Of all, the metric system of units was a more coherent system of units since it was easy to use and one may not use extra conversion factors that are not available in the equation.

1.4.1 Metric system of units

The metric system is the most common and internationally accepted decimalized system of measurement also known as SI unit system. It is classified in terms of base units (fundamental) and derived units. These units can be used to define any other physical quantities. The metric system of units is easy to use with the help of multiples and fractions. In fact, the earlier used US Customary unit system and British Imperial unit system are defined using the metric unit system.

The very idea of introducing the metric system was to have a unit system derivable from an unchanging phenomenon but practical limitations made the use of artefacts (prototype for metre and kilogram) indispensable. The multiples and submultiples of metric units in terms of powers of ten make them conveniently useful with prefixes in the names. Metric system exists in various forms with the option of choosing different base units. CGS (centimetre–gram–second) and MKS (metre–kilogram–second) are also known as metric system. FPS (foot–pound–second) is known as British System. SI unit system is more precise form of MKS system. Metric units are widely used and accepted system of units in science.

Units are broadly classified as

(1) *Fundamental units*: Also known as base units, these units are the units of fundamental quantities which can be measured independently; for example, length, mass, time, electric current and candela.

(2) *Derived units*: These units are derived by use of two or more fundamental units. For example, area of rectangle, volume, energy, force, velocity, pressure, lumen and watt.

All physical quantities can be expressed in terms of these seven base quantities. Table 1.2 lists the seven base units along with their standard definition and Table 1.3 represents some derived units using base units.

1.4.2 Classification of metric system of units

There are three types of fundamental unit systems in metric system.

1. **CGS unit system:** As the name indicates, this metric system uses centimetre, gram and second as the base units of length, mass and time.

2. **MKS and MTS system:** This system uses metre, kilogram and second as base units. It was adopted in 1889 and it replaced CGS unit system. This MKS system formed the basis of SI. There is another system of units called MTS system based on metre, tonne and second.

3. **SI unit system:** Earlier, different unit systems were used to measure the same physical quantity, but then a standard procedure was developed to maintain uniformity in the measurement called System de International (SI) unit system. It was introduced by the International Union of Pure and Applied Chemistry (IUPAC) and International Organization for Standardization, also known as SI. It is a type of metric system developed in 1960 when some other units were also incorporated, which were earlier absent in the other metric system. It is the most widely accepted unit system. The SI unit system is derived from MKS system.

Table 1.2: Base units for seven physical quantities with SI units.

Physical quantity	Symbol	Name of SI unit	Symbol for SI unit
Length	L	Metre	m
Mass	M	Kilogram	kg
Time	t	Second	s
Electric current	I	Ampere	A

Table 1.2 (continued)

Physical quantity	Symbol	Name of SI unit	Symbol for SI unit
Temperature	T	Kelvin	K
Amount of substance	N	Mole	mol
Luminous intensity	I_v	Candela	cd

British Imperial Unit system
The imperial system or imperial units is a system of units defined in British Weights and Measurements Act of 1824. In British imperial system, the base unit of length is yard and that of temperature is Fahrenheit. It was derived haphazardly but later by twentieth century most countries adopted the metric system (Appendix A1). FPS is the fundamental unit of British unit system which is foot–pound–second.

US Customary System of Units
The other type of traditional unit system that was established in the United States was the US Customary System of Units that was derived from the imperial system of units; for example, feet, inches, miles and yard (Appendix A1). The United States after its independence from Britain decided to adopt only the imperial system of units, even when the metric system was becoming popular among most countries.

1.5 Definition of base units

The base units are defined in terms of constants like speed of light in vacuum (c), charge of the electron or those that can be measured accurately. The new redefined units have replaced the artefact unit system with more coherent units of measurements. The earlier used prototype of weight (platinum–iridium alloy) did not always weigh the same. Even when the piece was inside three bell jars, it gets dirty and rusty. Cleaning the dust would change the mass of the prototype. Hence, it was not accurate. The redefined system allows to measure mass directly at any scale and will prove to be a boon to the scientists.

The following physical constants have been redefined in SI unit system:
(1) The Planck's constant
(2) The Avogadro's constant (N_A)
(3) The Boltzmann's constant (k)
(4) The elementary electric charge (e)
(5) The speed of light (c)
(6) The luminous efficacy of a defined visible radiation (K_{cd})
(7) The caesium hyperfine frequency (Δv_{cs})

In November 2018, the changes in the units for four units were approved by General Conference on Weights and Measurements. In May 2019, four of the seven SI base units were redefined, which are kilogram, ampere, kelvin and mole by setting the exact value to the Planck's constant (h), Avogadro's constant (N_A), Boltzmann's constant (k) and the elementary electric charge (e), respectively. The other three units, that is, second, metre and candela, were already defined by the physical constants. All the seven base units that are redefined can be explained further.

(1) **Length**: Its base unit is metre. Earlier, the definition of metre was based on wavelength of spectral line of krypton-86 and was defined as the distance travelled by light in a vacuum over a time interval of 1/299,792,458 of a second. The current definition is based on taking fixed numerical value of speed of light, c, in vacuum to be 299,792,458 m s^{-1}.

(2) **Mass**: The base unit of mass is kilogram (kg). Earlier, it was defined as the mass of the shiny piece of platinum–iridium alloy held in Bureau International des Poids et Mesures in France. It used to be the international prototype or artefact of the kilogram. The new definition of kilogram involves accurate weighing machines called "kibble balance" that uses Planck's constant to measure mass precisely by using electromagnetic force. The Planck's constant which has been measured in recent years with precision is $6.62607015 \times 10^{-34}$ kg m^2 s^{-1}.

This new SI system will bring more accuracy in many fields, including biotechnology, manufacturing, trade and human health.

(3) **Time**: The base unit for time is second. A second is defined as the duration of 9,192,631,770 periods of the radiation corresponding to the transition between two hyperfine levels of the ground state of the caesium-133 atom. In the improved definition, the numerical value of the caesium frequency Δv_{cs} is fixed.

(4) **Current**: The base unit for current is ampere. Ampere is defined as the amount of electric charge passing through an electric circuit per unit time but the prevailing definition defined it as the flow of $1/1.602176634 \times 10^{-19}$ times the elementary charge, e, per second.

(5) **Temperature**: The base unit of temperature is Kelvin. The previous definition of kelvin was defined as the fraction 1/273.16 of the thermodynamic temperature of the triple point of water, that is, 0.01 °C or 273.16 K while the current definition involves fixing the numerical value of Boltzmann's constant, k, to be 1.38×10^{-23} J K^{-1} making sure that the triple point of water temperature remains the same. One kelvin is equal to change in temperature T that results in thermal energy kT by 1.380649×10^{-23} JK^{-1} where J K^{-1} = kg m^2 s^{-2} K^{-1} where kilogram, metre and second are defined according to the improved definition. Now, since kelvin is derived from physical constant rather than experimental condition (triple point of water), it is more reliable and stable.

(6) **Amount of substance**: The base unit of amount is mole. It was defined as the number of atoms in 12 g of carbon ^{12}C (the isotope of carbon with relative atomic mass

12 Da). The revised definition is the amount of substance of exactly $6.02214076 \times 10^{23}$ elementary particles (atoms, molecules or ions).

(7) **Luminous intensity**: Its base unit is candela (cd). It is defined as the luminous intensity or electromagnetic field of frequency 540×10^{12} Hz or 540 terahertz emitted by a source in a specified direction with 1/683 watt per steradian. Luminous intensity measures the visual intensity of any light source and is the only unit that a human eye can measure. It reflects the brightness or intensity of the light and its direction. Since earlier times, candles were used to be the source of light; hence, the name is derived from the candle, so candela being the unit of luminous intensity.

Table 1.3: Derived SI units.

Physical quantity	Name	Description	Symbol	SI units
Frequency	Hertz	Events per unit time	Hz	s^{-1}
Force	Newton	Mass $\times$ acceleration	N	$kg\ ms^{-2}$
Pressure	Pascal	Force per unit area	Pa	$N\ m^{-2}$
Energy, work, heat	Joule	Force $\times$ distance	J	N m
Power	Watt	Work per unit time	W	$J\ s^{-1}$
Electric charge	Coulomb	Current $\times$ time	C	A s
Electric potential	Volt	Work per unit charge	V	$J\ C^{-1}$
Electric capacitance	Farad	Charge per unit potential	F	$C\ V^{-1}$
Electric resistance	Ohm	Potential per unit current		$V\ A^{-1}$
Electric conductance	Siemens	Current per unit potential	S	
Magnetic flux	Weber	Work per unit current	Wb	$J\ A^{-1}$
Magnetic flux density	Tesla	Magnetic flux per unit area	T	$Wb\ m^{-2}$
Inductance	Henry	Magnetic flux per unit current	H	$Wb\ A^{-1}$
Plane angle	Radian	Angle subtended by unit arc at centre of unit circle	rad	1
Solid angle	Steradian	Solid angle subtended by unit surface at the centre of unit sphere	Sr	1

Some physical quantities are dimensionless since they have the ratio of quantities with same dimension. For example, mole fraction, relative pressure, relative mass and so on. Practically, there is no considerable effect on daily life measurement. It will only matter if the measurements are below 1 kg or while measuring very small mass. The only advantage now is that the scientist would now be able to measure them at any place, time or scale. Hence, more accurate measurements would be made.

1.6 Interconversion of units

There are some experiments where the measurement of the physical quantity in one type of unit is more preferable than in other. The process of conversion depends on the situation and intent of purpose for which conversion is required. For instance, the SI unit for potential difference is volt, but while performing potentiometric titrations, the preferred unit is mV since the amount of potential difference is very small. In such cases, one needs to carry out the interconversion of unit into more accepted form, that is, SI unit. Similarly, if one wants to measure the length of a pencil, then some measurements if made in metre (SI unit) would introduce more error and approximation, rather if the measurement is made in cm (CGS unit), then converting into metre would give more accurate result. There are many multiples and submultiples of metre, some are millimetre, micrometre, centimetre, kilometre and so on. They have prefixes attached to the SI units (Appendix A1). On the basis of utility, one may use the conversion factor to convert the measured unit into more preferred unit. A conversion also depends on the precision and accuracy of measurement with which comes certain amount of associated uncertainty of measurement. Conversion should neither increase nor decrease the precision of the first measurement. A table of different relationships and conversion factors for some units is given at the end of the book (Appendix A1). The conversion from one unit form to another unit form can be done by simple method called **factor label method**. Factor label method is an elementary method used to convert a quantity measured in one unit to another unit system by multiplying the conversion factor.

Example 1: Convert 0.07912 metre (m) into millimetre (mm).
Solution: From the conversion factors, we know that 1 m = 1,000 mm

$$\text{So } 0.07912\,m = 0.07912\,m \times \frac{1,000\,mm}{1\,m}$$

$$0.07912\,m = 79.12\,mm$$

When the physical quantity is multiplied by the conversion factor ratio, then one unit in the denominator is cancelled out by unit in numerator leaving only one unit.

Example 2: A piece of metal is 7 in long. Calculate its length in cm.
Solution: Since 1 in = 2.54 cm

$$\text{Therefore, } 7 \text{ in} = 7 \text{ in} \times \left[\frac{2.54 \text{ cm}}{1 \text{ in}}\right] = 17.78 \text{ cm}$$

Example 3: A beaker contains 5 L of NaOH. Calculate the volume of NaOH in m^3.
Solution: As 1 L = 1,000 cm^3 and 1 m = 100 cm

$$\text{Hence, } 5 \text{ L} = 5 \times 1,000 \text{ cm}^3 \times \left[\frac{1 \text{ m}}{100 \text{ cm}}\right]^3 = 5 \times 10^{-3} \text{ m}^3$$

Example 4: How many atoms are there in 7.50 g of aluminium coil?
Solution: The atomic mass of aluminium is 26.981539 g/mol, which implies that 1 mol of aluminium contains Avogadro's number (6.023×10^{23}) of aluminium atoms or 26.981539 g of aluminium contains 6.023×10^{23} aluminium atoms. So, by using factor label method, we have

$$7.50 \text{g} = 7.50 \text{g} \times \frac{\text{mol}}{26.981539 \text{g mol}^{-1}} \times \frac{6.023 \times 10^{23} \text{atoms}}{\text{mol}}$$

$$= \frac{7.50 \times 6.023 \times 10^{23}}{26.981539} \text{ atoms}$$

$$= 1.67 \times 10^{23} \text{atoms}$$

1.7 Significant figures

While making measurements, the instrument measures the physical quantity which is accurate up to a certain limit. The digits in the measurement up to which the measurement is certain are called significant figures. The number of digits implies the certainty or confidence up to which a measurement is made. More significant digits imply greater precision.

Certain rules to be followed to determine the significant figure in any measurements are listed as follows.

1. *All non-zeroes in a measurement are significant.*
   For example, mass of a metal piece is 3.15 g. Here, there are three significant figures.
2. *All the zeroes between two non-zero digits are significant.*
   For example, in numbers 0.00000208, 0.2008 and 208.00, there are three, four and five significant numbers, respectively.
3. *A measurement which is less than one, all zeros to the right of a decimal point and to the left of a non-zero digit are never significant.*
   For example, in 0.0234 g, number of zero after decimal and before 2 is not significant. There are three significant digits: 2, 3 and 4.

4. **(a) All zeroes on the right of the last non-zero digit with the decimal point are significant.**
   For example, in a number "2350.", there are four significant digits.
   (b) All zeroes on the right of the last non-zero digit with no decimal point are insignificant.
   For example, in the number 2350, there are three significant figures.
   (c) The trailing zeroes to the right of a decimal point are significant, if they are not followed by a non-zero digit.
   For example, in number 53.00, there are four significant digits (5, 3, 0 and 0)
5. **The exact numbers have infinite number of significant figures.**
   For example, 3 m = 3.00 m = 3.0000 m = 3.00000000000 m
6. **When the number is written in scientific notation as $a \times b^n$, then the number of significant figures are counted in a.**
   For example, we have the number 0.000521, which we can write in scientific notation as 5.21×10^{-4}. Therefore, here the number of significant digit is three (5, 2 and 1). The 10 and –4 do not hold any place in significant figures. Changing any number to scientific notation may vary with the number of significant figures.
   For example: 3600
   3600 = 2 s.f.
   3600. = 4 s.f.
   3.60×10^3 = 3 s.f.

Atlantic–Pacific rule

If a number has a decimal point, then the trailing zeroes on the Pacific/left side are insignificant, and if the number does not have a decimal point, then the zeroes to the Atlantic/right side are insignificant (since Pacific ocean is on left and Atlantic is on right side of the United States).

For example, 0.00837 has 3 s.f. (8, 3 and 7) while 3500 has 2 s.f. (3 and 5).

1.8 Rules for performing arithmetic operations

1.8.1 Addition and subtraction

The significant figure while performing addition and subtraction depends on the number of decimal places as the number with least precise value.

For example, 63.152 + 5.6 = 68.752

Instead, one should write the answer as 68.7, since the least precise value (5.6) has one decimal place.

Example 5: $X = 312.951 + 0.32$.
Solution: $X = 313.27$
  Since, the least precise term in the addition is 0.32, which has two decimal places. So, the answer should also retain two decimal places. Hence, it is round off to two decimal places, that is, 313.27.

Example 6: 0.241 g of NaOH is dissolved in 142.0 g of water. What is the mass of the solution?
Solution: We can find the mass of solution as 142.0 g + 0.241 g = 142.241 g.
  Since the least precise measurement has 1 decimal place (i.e. 142.0), then the answer should be 142.2 g (by rounding off).

1.8.2 Multiplication and division

The same rule applies while performing multiplication and division, that is, the number of significant figures in the result would be the same as the least precise number in the calculation.

Example 7: $1.5 \times 2.46 = 3.69$ (3 s.f.).
Solution: 1.5 has 2 s.f. and 2.46 has 3 s.f., so the answer according to the rules should have as many s.f. as in the least precise number which is 1.5.
  Hence, the result on rounding off should give 3.7 (2 s.f.).

Example 8: $1.34 \times 5.073 \times 3.7 = 25.151934$
Solution: Since the least precise number is 3.7 which has 2 s.f., the result should be rounded off to 25, which has 2 s.f.

Example 9: 1.50×6.234
Solution: Each of the above has 3 and 4 s.f., respectively. The answer should be 9.351. Again applying the rule, the rounding off should give 9.35(3 s.f.).

Example 10: 19.3 mL of 0.05721 N NaOH was used to neutralize the given 20.0 mL HCl solution. Calculate the normality of HCl solution used.
Solution: Here the normality of NaOH has 4 s.f. but the other volume values have 3 s.f., so the final answer should also have least number of s.f., that is, 3. We know that $N_1V_1 = N_2V_2$ so

$$N_{HCl} = 19.3 \times 0.05721/20.0$$
$$= 0.05520$$
$$= 0.0552 (\text{rounding off})$$

While calculating the logarithm, the mantissa of the logarithm should contain the same number of significant digits as there are in the number for which the logarithm is calculated.

For example, $\log(2.00 \times 10^5) = 5.3010299956$. Since the number 2.00×10^5 has 3 s.f., the mantissa should also have 3 s.f. Hence, the answer should be 5.301.

For example, $\log(2 \times 10^4) = 4.3010299956$. Here, 2×10^4 has 1 s.f. so answer should be 4.3.

Example 11: The rate constant at two temperatures 289 and 333 K are 5.03×10^{-2} and 6.71 mol^{-1}dm^3 s^{-1}, respectively. Calculate the value of activation energy using the expression

$$\ln\frac{k_{333}}{k_{298}} = \frac{E_a}{R}\left(\frac{1}{289} - \frac{1}{333}\right) \tag{1.8}$$

Solution: Both the constants have 3 s.f., so the ratio of k_2/k_1 should also have 3 s.f. as $k_2/k_1 = 133.39$. Since the rule says that the mantissa should have as much s.f. as in the number itself for which the logarithm is calculated so the number of s.f. should be 3 in mantissa as $\ln k = 4.893$, which gives $E_a = 88,982.709$ Jmol^{-1}

While calculating antilog of a number, the result should have the same number of s.f. as the mantissa in the logarithm.

For example, antilog(0.477) = 2.99916251, which should be rounded off to 3.00

For example, antilog(0.47) = 2.951209226, which should be rounded off to 3.0

Example 12: The Arrhenius equation is given as

$$\ln k = \ln A - E_a/RT \tag{1.9}$$

Given $\ln A = 9.874$ and $E_a = 28.26$ kJ mol^{-1}. Calculate k at 298.15 K.

Solution: $\ln k = 9.874 - (28.26 \times 10^3)/(8.314 \times 298.15) = 9.874 - 11.40 = -1.5266$

The mantissa has only one significant figure. Then solving for k and using the above-mentioned rule, we have

$k = e^{-1.5266} = 0.217$ or just 0.2.

1.8.3 Mixed mathematical operations

When mixed operations are involved, calculation should be done using more number of significant figures than will be significant to arrive at a value. Then, go back and retrace each step for how many significant figures should have been left up to the final result based on the standard convention. Successive rounding can magnify the inaccuracies. Hence, when doing rounding at each step, it is always advisable to retain one extra insignificant figure than can be justified to avoid error. The final answer is then rounded off to the desired number of significant figures. For instance,

$$X = ((4.354 + 0.0026)/28.5) - 3.553 \times 10^{-3}$$

$$X = 0.14931016$$

First, 4.354 + 0.0026 = 4.357 (since the least precise value has accuracy of 3 decimal places so should be in the answer (instead of 4.3566)).

Second, 4.357/28.5 = 0.153 (since the least precise number has 3 significant figures instead of 0.152877192).

Lastly, 0.153 − 0.003553 = 0.149 (since in subtraction, the least precise number has three decimal places, that is, 0.153 (instead of 0.149447)).

The value obtained while keeping all figures was 0.14931016 which if rounded off to have three significant figures is 0.150 or 1.50×10^{-1}, which is the final answer.

1.9 Rounding off significant figures

Rounding off the numbers is done to remove insignificant figures by replacing them with insignificant zeroes. If the insignificant digit is 0, 1, 2, 3 and 4, then while rounding off, one must underestimate the value. The value of the last significant figure remains unchanged. If it is 5, 6, 7, 8 and 9, then overestimation is done and 1 is added to the last significant digit.

Example 13: Round off 2.568 to 4 s.f.
Solution: 2.568 is rounded off to 2.57 (3 s.f.). In this case, the insignificant number is greater than 5 then it is easy to round it off to the next highest number, but if the insignificant digit is 5, there are two options, either round up or round down because in that case the number is lying between two rounded numbers.

Example 14: Round off 5321.98 to 2 s.f.
Solution: 5300, since it has no decimal point, the number of significant figures is 2. If it had been 5300., then it will have 4 s.f., which is not correct.

Example 15: Round off the following:
(a) 2.3467 to 4 s.f.   (b) 3.6541 to 3 s.f.
Solution: (a) In 2.3467 the last significant digit is 6 and insignificant digit is 7. Since it is greater than 5, 1 is added to the last significant digit and the insignificant digit is dropped off.

(b) 3.6541 is round off to 3.654. Here, the last significant digit is 4 while insignificant is 1 which is less than 5; hence, it should be dropped off and only the previous digits are kept.

There is one more method to round off. It revolves around rounding off around the digit 5. This value is midway between underestimated and overestimated value. In such cases, 5 is rounded down when the preceding significant digit is even and 5 is rounded up when the significant digit preceding is odd.

Example 16: 2.6375 is to be rounded off to 4 s.f.
Solution: Here, the last insignificant figure is 5 while the significant figure is 7, which is odd; hence, the rounding off gives 2.638 to make the last significant digit even (7 to 8).

Example 17: 2.3145 is to be rounded off to 3 s.f.
Solution: Here the last insignificant figure is 5 while the significant figure is even (4) already. Hence, it should not be meddled with. Hence the answer should be 2.314 (4 s.f.).

**Insignificant figures are often written as a small size letter like 2.345$_7$, where 7 is the insignificant figure and 5 is the last significant figure.

1.10 Problems for practice

1. Identify the number of significant figures:
   (a) 2.0040
   (b) 0.083
   (c) 5.21×10^{-5}
   (d) 9,600
   (e) 7,500.00
   (f) 9.80×10^{7}
   (g) 6.023×10^{23}

2. Round off the following numbers:
   (a) 6,509,213.246 to 3 s.f.
   (b) 0.0672000 to 1 s.f.
   (c) 92.3518 to 3 s.f.
   (d) 0.0012475 to 3 s.f.
   (e) 0.004520 to 2 s.f.
   (f) 5.0942 to 2 s.f.
   (g) 32,391 to (1 s.f. to 6 s.f.)
   (h) 83.810482 to (1 s.f., 2 s.f. and 6 s.f.)
   (i) 0.7615 to 2 s.f.
   (j) 0.956 to 1 s.f.
   (k) 894.2 to 1 s.f.
   (l) 8.2197 to 1 s.f.
   (m) 39.0126 to 1 s.f.
   (n) 0.0247779 to 4 s.f.
   (o) 8,516.131 to 2 s.f.
   (p) 0.00031834662 to 2 s.f.

3. Find the following:
   (a) $X = 62.4 + 19.570$ to 3 s.f.
   (b) $Y = 0.07215 - 0.0002138$ to 4 s.f.
   (c) $Z = 98.1 \times 0.03$ to 1 s.f.
   (d) $W = 9.568/4.61457$ to 4 s.f.

(e) $V = 3.862 \times 0.62$ to 2 s.f.

(f) $U = 82.5/2.3186$ to 3 s.f.

(g) $K = 712368 - 612368$ to 1 s.f.

(h) $A = (1.2673 \times 10^7)/(3.95 \times 10^{-4})$

(i) $F = (14.5 \times 12) - (35.6/6.09)$

(j) $L = ((8.325 \times 10^3) \times (3.1729 \times 10^{-7}))/(3.9641 \times (7.2126 \times 10^{-5}))$ to 4 s.f.

Chapter 2
Uncertainties and errors

2.1 Introduction

The physical quantity measurements are always accompanied by error since nothing can ever be measured with certainty. This uncertainty is referred to as Error. Error is a mathematical representation of the uncertainty. Error should not be confused with mistake. Mistake can be corrected (like mistake during calculation or solution preparation or wrong reading of the value, etc.). Error analysis facilitates the identification, quantification and elimination of error and makes the methods of experiment more refined and robust. During approximation or rounding off, a large amount of error is introduced. Hence, knowing the nature of error as well as the quantification of error helps to interpret the method in a more robust manner. Moreover, if the errors are more than allowed limits, then we would know whether to keep or get rid of the particular method or observations. This error analysis is more important when we talk about numerical methods and computational calculations that require more accuracy and precision.

Errors are broadly classified into two categories as systematic errors (determinate errors) and random errors (indeterminate errors). Systematic errors can be identified and corrected while random errors cannot be eliminated.

Uncertainty refers to the doubt around the measurement. Even with well-calibrated instruments (thermometer, rulers, clocks, etc.), when measurements are taken, there is always a margin of doubt. Uncertainty is the range of possible values that any physical quantity can take during measurement. When experimental results are reported, the true value is not reported but a range of values are reported and then we should find if the accepted value falls under the range of uncertainty values.

There are many factors that influence the measurements including temperature, pressure, humidity and calibration. When multiple measurements are made for the same physical quantity, then a slightly different measurement value is obtained each time. In fact, uncertainty adds reliability to the results in a particular range. To quantify the doubt, one must know two terms, namely, the interval and the confidence limit. The interval depicts the width of margin or doubt, while confidence level indicates how sure the true value is within that interval or margin. The best way to show the uncertainty is

$$\text{Measurement} = \text{best estimate} \pm \text{uncertainty}$$

For instance, a copper wire is 3 m long with an uncertainty of 15 cm (0.15 m), which can be represented as

$$\text{length} = 3.0 \ \pm \ 0.15 \,\text{m}$$

https://doi.org/10.1515/9783111334448-002

Here the length of copper wire is reported as 3 ± 0.15 m at the 95% confidence level. It implies that there is 95% certainty that copper wire is between 2.85 and 3.15 m long. In fact when in laboratory, the students take the readings, they do not calculate the percent error, but the percent uncertainty. The error is calculated when the true value is known and it is almost next to impossible to know the actual true value.

Uncertainty is important as one wants to improve the quality of measurements. This is why while making calibration or performing a test, the uncertainty must be mentioned as a measure of success or failure. To avoid uncertainty in measurements, the measurement should be taken at least thrice. So, the error that would have gone unnoticed in the first measurement may be noticed in the next two measurements.

To further corroborate the authenticity of measurements, statistical calculations may be carried out on the measurements to account for the error and uncertainty in the data. So, if a set of measurements is taken, then average of all the measurements will give the estimate of true value. More number of readings will give better average, which would be more close to the true value. When the repeated measurements are different, there is a lot of uncertainty involved for that physical quantity, (the spread is large) but there is a possibility that average of all the measurement will give a best estimate of the true value. A large spread implies that there is a huge difference between the set of readings for the same experiment. The standard way to quantify spread is standard deviation. It tells how an observation deviates from the average or mean value. Standard deviation is often greater than average deviation and is mostly used by statisticians and scientists.

Consider weighing a piece of copper alloy, where the weighing was repeated thrice and each time a different mass was obtained, for example, 1.213, 1.275 and 1.165 g, respectively. The average for the above measurements is 1.217:

$$\text{Average} = \frac{1.213 + 1.275 + 1.165}{3} = 1.217 \tag{2.1}$$

Average deviation is often calculated to display results as

$$|1.217 - 1.213| = 0.004 \tag{2.2}$$

$$|1.217 - 1.275| = 0.058 \tag{2.3}$$

$$|1.217 - 1.165| = 0.052 \tag{2.4}$$

Average deviation is calculated to be 0.038:

$$\text{Average deviation} = \frac{0.004 + 0.058 + 0.052}{3} = 0.038 \tag{2.5}$$

Hence,

$$\text{Percent average deviation} = \frac{0.038}{1.217} \times 100 = 3.12\% \tag{2.6}$$

Hence, 3.12% is the percent average deviation but often standard deviation is used to depict uncertainties. As standard deviation is

$$s = \sqrt{\frac{\sum_{i=1}^{n} (x_i - \bar{x})^2}{n-1}} \tag{2.7}$$

$$s = \sqrt{\frac{(0.004)^2 + (0.058)^2 + (0.052)^2}{3-1}} \tag{2.8}$$

$$s = 0.055$$

Therefore, 5.5% is the percent standard deviation, which is a more standard and common practice to denote deviations.

Systematic errors

Systematic errors are the errors that can be identified for their source and, hence, can be eliminated. Systematic errors often lead to measured value, either very high or very low. They are also called "zero error", which may be positive or negative. Further, they can be classified into four types:

(1) **Instrumental error**: This type of error originates due to poor calibration, faulty part of instrument or electricity fluctuation. A poorly calibrated instrument may show zero error due to environmental factor which can be corrected by calibrating the instrument. For example, a poorly calibrated thermometer when immersed in boiling water reads 103 °C. Different pipettes and volumetric flasks hold different volumes. Hence, such type of error will only magnify the error in further readings.

(2) **Observational error or personal error**: These types of errors arise due to personal drawback of observation. These errors may include mishandling of experimental apparatus, improper and careless weighing, vagueness in taking readings, poor judgement or prejudice or color ambiguity including parallax in reading a metre scale, reading upper or lower meniscus. It can be corrected by proper training and experience.

(3) **Theoretical error**: For the sake of simplification, if one approximates a condition in the model system, then such error arises. For example, if one assumes atmospheric pressure does not change the readings of temperature, then it will definitely introduce the error in the readings.

(4) **Methodic error**: This kind of error originates when the system, either chemical or physical, behaves erratically. For example, the occurrence of side reaction (e.g. co-precipitation), volatility, insolubility of precipitates, decomposition of species, unexpected rate of reaction (slow or fast), instability of reaction species and so on. It is the most serious of all errors. It can be corrected by developing proper methods.

All the above systematic errors once identified can be eliminated and the results will improve.

Random errors

Random errors are also called as indeterminate errors They may occur due to sudden change in experimental conditions like change in temperature, humidity, current and so on.The sources of such errors are unidentified and, hence, they are difficult or impossible to eliminate. They may also occur due to the inability of analyst to reproduce the same readings. Random errors follow Gaussian distribution. For example, if we have measured the weight of a ball four times using the same measuring scale and it turns out to be 20.23, 20.25, 20.27 and 20.26 g. So, to avoid this error, more number of readings could be taken and then an average of all observation is calculated.

2.2 Absolute and relative uncertainty

Uncertainty can be classified as absolute and relative uncertainty similar to absolute and relative error. Absolute uncertainty reflects the margin of uncertainty in a measurement while relative uncertainty is the ratio of absolute uncertainty and the size of its associated measurement. Relative uncertainty alike relative error is dimensionless.

Relative uncertainty = absolute uncertainty/magnitude of measurement

Percent relative uncertainty = 100 × relative uncertainty

Here, it should be pointed out that accuracy is not the same as certainty. While accuracy is qualitative, uncertainty is quantitative. The plus–minus sign is used when discussing errors and uncertainties.

Accuracy and precision
Accuracy reveals how closely observed value is to the true value, while precision reveals how closely different observation values agree with each other. Precision is also known as reproducibility. One may be precise and inaccurate and vice versa. In sciences, accuracy and precision are integral to measurements and calculations. While measuring, there is always an error associated with each value, precision measures the extent or spread of error from the real value. It is independent of true actual value. Precision is expressed in digits (for measurement) but can also be expressed in terms of deviation of the errors. A three digit measurement is more precise than two digit measurement.

For example, suppose there are six readings of marks obtained by students in a test as 8.1, 8.2, 8.5, 8.2, 8.2 and 8.5, if they are wrongly uploaded on the web as 3.1, 3.2, 3.5, 3.2, 3.2 and 3.5.

Then, the marks that have been recorded would still be precise but not accurate. Accuracy is reduced drastically here since the actual marks and uploaded marks have great differences.

Likewise, error should not be confused with uncertainty. While error is the difference between the true value and measured value, uncertainty is the quantification of doubt around the measurement. Any error due to known sources could be corrected but when the source, is unknown, such errors are said to be uncertainty. One may never make a perfect measurement which gives a true value of a physical quantity; hence, measurements are often estimates of true value. Therefore, what scientists are interested in experiments is uncertainty.

2.3 Propagation of uncertainty

Systematic errors can be corrected if one knows the source while random errors cannot be corrected. Since calculations involve numbers (of physical quantity) that have certain degree of uncertainty (random error) associated with them, the uncertainties can also be combined if they have the same units. For example, while measuring mass, the uncertainty should also be mentioned in units of mass. This procedure of calculating one quantity using multiple quantities is also called data reduction. These calculations involve basic arithmetic operations like addition, subtraction, division and multiplication. This uncertainty may then be carried forward to the final result of the calculations and hence is called propagation of uncertainty. These are also called *combining uncertainties*.

The propagation of uncertainties involving arithmetic operations follow these simple rules.

(1) Addition and subtraction

While performing addition and subtraction, if e_1, e_2 and e_3 are the uncertainties associated with three measurements as $x \pm e_1$, $y \pm e_2$ and $z \pm e_3$, then the resultant absolute uncertainty e is given as

$$e = \sqrt{e_1^2 + e_2^2 + e_3^2} \tag{2.9}$$

Example 1: The three values of a, b and c are given as 2.16 ± 0.02, 1.46 ± 0.15 and 4.37 ± 0.04, respectively. Evaluate variable d which is given as

$$d = a - b + c \tag{2.10}$$

Solution: $d = 2.16 - 1.46 + 4.37 \pm e = 5.07 \pm e$

5.07 is the arithmetic answer and e is the uncertainty associated with the result, which can be deduced using eq. (2.9):

$$e = \sqrt{(0.02)^2 + (0.15)^2 + (0.04)^2} \tag{2.11}$$

$$e = 0.15$$

This is known as absolute uncertainty. Hence, the final result would be $d = 5.07 \pm 0.15$.

Likewise, percent uncertainty can be calculated as

$$\text{Percent uncertainty} = \frac{0.15}{5.07} \times 100 \tag{2.12}$$

$$\text{Percent uncertainty} = 3.0\%$$

Hence, the result in terms of percent uncertainty is 5.07(±3.0%).

**Apart from the significant figure, an extra insignificant figure (as subscript) is kept in the uncertainty to avoid the round-off error later in the calculations.

Example 2: The initial and final readings of a burette during a titration are 16.85(±0.03) mL and 0.03 (±0.03) mL, respectively. What is the uncertainty in the volume delivered?

Solution: As volume delivered is the difference between final reading and initial reading, the volume transferred is

$$16.85(\pm 0.03) - 0.03(\pm 0.03) = 16.82(\pm e) \tag{2.13}$$

where

$$e = \sqrt{(0.03)^2 + (0.03)^2} \tag{2.14}$$

$$e = 0.04_2$$

Here, although the uncertainty associated with each reading is 0.03, the resultant uncertainty in delivering the volume is 0.04.

How many digits should be kept?

(a) Uncertainties are mostly quoted to one significant digit. For example, 0.02 rather than 0.023. Only if the uncertainty starts with 1, then in such cases uncertainty is quoted to two significant digits. For example,

23.5 ± 0.23	wrong	23.5 ± 0.2	correct
13.7 ± 0.1	wrong	13.7 ± 0.15	correct
98.7 ± 3.2	wrong	98.7 ± 3	correct

(b) Also, the experimental measurements are round off to the same decimal place as the uncertainty. For example

45.378 ± 0.1	wrong	45.3 ± 0.1	correct

(c) The measurement uncertainties should overlap. Only if the difference is not very large, the measurements are considered real; else, they are considered to be sloppy observations.

For example, the measurements 0.56 ± 0.02 and 0.68 ± 0.02 do not agree since their spread does not overlap at all (maximum values and minimum values for two measurements are (0.58 & 0.54) and (0.70 & 0.66), respectively), while the measurements 0.56 ± 0.06 and 0.68 ± 0.07 do overlap. Here the second set measurement is real and acceptable.

(d) The result should have same number of significant figures as that of the quantity which is less precise. For instance, area of circle is given by $A = \pi r^2$, where π has many number of significant figures since its value is known up to 3.1415927..., while r is 2.5 cm; hence, the result would depend on less precise quantity, which is r having only 3 s.f.

---

(2) Multiplication and division

While performing arithmetic operations like multiplication and division, the absolute uncertainties should first be converted into percent uncertainty and then uncertainty is defined in terms of percent uncertainty as

$$e = \sqrt{(\%e_1)^2 + (\%e_2)^2} \tag{2.15}$$

Example 3: Evaluate

$$d = \frac{3.36 \pm 0.04}{1.07 \pm 0.02} \tag{2.16}$$

Solution: The above operation may be written in terms of percent uncertainty as

$$d = \frac{3.36 \pm 1.2\%}{1.07 \pm 1.8\%} \tag{2.17}$$

$$d = (3.36/1.07) \pm e \tag{2.18}$$

$$d = 3.14 \pm e \tag{2.19}$$

where

$$e = \sqrt{(\%e_1)^2 + (\%e_2)^2} \tag{2.20}$$

or

$$\frac{\Delta e}{e} = \sqrt{\left(\frac{\Delta e_1}{e_1}\right)^2 + \left(\frac{\Delta e_2}{e_2}\right)^2} \tag{2.21}$$

$$e = \sqrt{(1.2)^2 + (1.8)^2} \tag{2.22}$$

$$e = 2.1\%$$

Hence,

$$d = 3.14 \pm 2.1\% \text{ (in percent uncertainty)}$$

To convert back into absolute uncertainty, multiply percent uncertainty with the result as $3.14 \times 2.1\% = 0.065 \approx 0.07$

$$d = 3.14 \pm 0.07 \text{ (in absolute uncertainty)}$$

(3) Mixed operations

When given a combination of arithmetic operations like addition and multiplication then in such cases, first the addition and subtraction operations are performed if any and then multiplication and division.

Example 3: Calculate d, where $a = 3.76 \pm 0.02$, $b = 1.42 \pm 0.01$ and $c = 2.03 \pm 0.04$

$$d = \frac{a - b}{c} \tag{2.23}$$

Solution: Accordingly, d should be

$$d = \frac{(3.76 \pm 0.02) - (1.42 \pm 0.01)}{2.03 \pm 0.04} \tag{2.24}$$

First, the numerator is evaluated, which turned out to be

$$(3.76 - 1.42) \pm e = 2.34 \pm e$$

where

$$e = \sqrt{(0.02)^2 + (0.01)^2}$$

(2.25)

Hence, numerator becomes   2.34 ± 0.02 (in absolute uncertainty)

2.34 ± 0.9% (in percent uncertainty)

So, now

$$d = \frac{2.34 \pm 0.02}{2.03 \pm 0.04}$$

(2.26)

$$d = \frac{2.34 \pm 0.9\%}{2.03 \pm 2.0\%}$$

(2.27)

$$d = 1.15 \pm e$$

where

$$e = \sqrt{(0.9)^2 + (2.0)^2}$$

(2.28)

$$d = 1.15 \pm 2.19\%$$

$$d = 1.15 \pm 0.025$$

Since the uncertainty is from 0.01 decimal places, the result should also be rounded off to 0.01 decimal place.

---

(4) Products of powers

If the variables involve exponents like the function as follows:

$$z = x^m y^n$$

(2.29)

In simple average terms, the uncertainty in z is given as

$$\frac{\Delta z}{z} = |m|\frac{\Delta x}{x} + |n|\frac{\Delta y}{y} + \cdots$$

(2.30)

If one uses the quadrature (root of sum of squares), then one may write as

$$\frac{\Delta z}{z} = \sqrt{\left(\frac{m\Delta x}{x}\right)^2 + \left(\frac{n\Delta y}{y}\right)^2 + \cdots}$$

(2.31)

Example 4: Evaluate d, when $a = 2.61 \pm 0.2$, $b = 3.7 \pm 0.36$ and $c = 1.8 \pm 0.65$, if

$$d = \frac{ab^3}{\sqrt{c}}$$

(2.32)

Solution: The arithmetic operation result would give

$$d = 98.539 \pm e$$

where e would be

$$\frac{e}{98.539} = \left(\frac{0.2}{2.61}\right) + 3\left(\frac{0.36}{3.7}\right) + 0.5\left(\frac{0.65}{1.8}\right) \tag{2.33}$$

which on solving gives

$$e = 54.105$$

Hence,

$$d = 98 \pm 54.$$

Since the uncertainty is very large, only two significant figures are kept and the terms after decimal point are ignored.

If the uncertainty in X^2 is to be considered, then it may be argued as $X \times X$ and rule of multiplication should be used but instead rule for powers of exponents is used. Since for multiplication rule, the variable should be independent of each other while measurements like X and Y, but in case of X^2, both are same.

(5) Multiplication by a constant

If there is any uncertainty in any measurable quantity and the quantity needs to be multiplied by a constant, then the uncertainty is also multiplied by the same constant.

Example 5: Calculate uncertainty associated with the perimeter of the square if the uncertainty in the side of square is 4.0 ± 0.3 cm.
Solution: As perimeter = $4 \times$ side, and considering the above-mentioned rule, the perimeter should be 16 ± 1.2 cm.

Example 6: Find z where $a = 3.0 \pm 0.2$, $b = 1.3 \pm 0.5$, $c = 5.2 \pm 0.02$

$$z = ab + c^3 \tag{2.34}$$

Solution: In this case, first the quantity ab is calculated with the errors, then c^3. These two quantities are then added separately with their errors.

ab is calculated first as

$$ab = 3.9 \pm 1.5$$

$$c^3 = 140.6 \pm 0.93$$

Hence,

$$z = 144.5 \pm 1.8$$

(6) Exponential and logarithms

For the function $y = x^a$, the percent uncertainty is given as

$$\%e_y = a\%e_x \tag{2.35}$$

Similarly for the function $y = \log x$, uncertainty in y is given as

$$e_y = \frac{1}{\ln 10}\frac{e_x}{x} \approx 0.4342\frac{e_x}{x} \tag{2.36}$$

Likewise, the other functions and the uncertainties associated with them are tabulated (Table 2.1).

Table 2.1: Functions and the uncertainty associated with them.

Function	Uncertainty	
$y = \ln x$	$e_y = \dfrac{e_x}{x}$	(2.37)
$y = 10^x$	$\dfrac{e_y}{y} = \ln 10\, e_x \approx 2.3026 e_x$	(2.38)
$y = e^x$	$\dfrac{e_y}{y} = e_x$	(2.39)

Example 7: In the function $y = \sqrt[3]{x}$, what is the uncertainty in y if the uncertainty in x is 3%.
Solution: Using eq. (2.35), the uncertainty in y would be

$$\%e_y = \frac{1}{3}(3\%) = 1\%$$

Example 8: Using the pH value of 4.18 ± 0.03, calculate the uncertainty in $[H^+]$
if

$$pH = -\log[H^+] \tag{2.40}$$

Solution: One may write eq. (2.40) as

$$10^{-pH} = [H^+] \tag{2.41}$$

Here, $y = [H^+]$, while $x = -pH$

$$y = 10^{-4.18} = 6.60 \times 10^{-5}$$

$$\frac{e_y}{y} = \ln 10\, e_x \approx 2.3026 e_x \tag{2.38}$$

$$\frac{e_y}{6.60 \times 10^{-5}} = 2.3026(0.03)$$

$$e_y = 0.45 \times 10^{-5}$$

Hence uncertainty in $[H^+] = 6.60(0.45) \times 10^{-5}$ M

Summation of quadrature

Combining uncertainties using standard deviation, one can express function f which is a function of x, y, z,... as $f(x, y, z,...)$, then the uncertainty in f can be written by taking partial derivative of f with respect to each variable. Then the total differential is

$$df(x, y, z, \ldots) = \left(\frac{\partial f}{\partial x}\right)dx + \left(\frac{\partial f}{\partial y}\right)dy + \left(\frac{\partial f}{\partial z}\right)dz + \ldots$$

We may write dx, dy, dz as the error in x, y, z,..., that is, Δx, Δy, Δz,...

$$\Delta f(x, y, z, \ldots) = \left|\frac{df}{dx}\right|\Delta x + \left|\frac{df}{dy}\right|\Delta y + \left|\frac{df}{dz}\right|\Delta z + \ldots$$

If we restrict ourselves to x and y only, then we can write

$$\Delta f^2 = \left(\frac{df}{dx}\right)^2 \Delta x^2 + \left(\frac{df}{dy}\right)^2 \Delta y^2 + 2\Delta x \Delta y \frac{df}{dx}\frac{df}{dy}$$

Since $\Delta x \Delta y$ are very small, we can write

$$\Delta f^2 = \left(\frac{df}{dx}\right)^2 \Delta x^2 + \left(\frac{df}{dy}\right)^2 \Delta y^2$$

If the standard deviations are small changes in x and y, then one may write the above equation as (standard deviation approach):

$$\sigma_f^2 = \left(\frac{df}{dx}\right)^2 \sigma_x^2 + \left(\frac{df}{dy}\right)^2 \sigma_y^2$$

$$\sigma_f = \sqrt{\left(\frac{df}{dx}\right)^2 \sigma_x^2 + \left(\frac{df}{dy}\right)^2 \sigma_y^2}$$

Hence, to sum up, in terms of standard deviation, the uncertainties are defined as

(1) When performing addition and subtraction, the resultant uncertainty σ_r can be written as

$$\sigma_r = \sqrt{\sigma_x^2 + \sigma_y^2 + \sigma_z^2}$$

where σ_x, σ_y and σ_z are the standard deviations associated with the variables.

(2) Similarly, when multiplication and division is involved, one may write the uncertainty as

$$\frac{\sigma_r}{r} = \sqrt{\left(\frac{\sigma_x}{x}\right)^2 + \left(\frac{\sigma_y}{y}\right)^2 + \left(\frac{\sigma_z}{z}\right)^2}$$

(3) Exponents $r = x^y$ as

$$\frac{\sigma_r}{r} = y\left(\frac{\sigma_x}{x}\right)$$

(4) Logarithmic function $r = \log(x)$ as

$$\sigma_r = 0.434\left(\frac{\sigma_x}{x}\right)$$

(5) Antilog $r = antilog(x)$ as

$$\frac{\sigma_r}{r} = 2.303\sigma_x$$

2.4 Rounding off uncertainties

Different calculators and spreadsheets may give answers to the calculation up to many decimal places. Sometimes the calculations are made simpler by rounding off the number up to some decimal places. If the uncertainty is in the first decimal place, then the measurement should also be reported in the first decimal place. For example,

$$3.0\,\text{m} \pm 0.15\text{m} \quad \text{correct} \quad\quad 3.02\,\text{m} \pm 0.15\text{m} \quad \text{wrong}$$

It is always advisable to make the calculations to at least one more significant figure than we require. Rounding off should be carried out at the end of the calculations to avoid rounding errors. If the rounding off is carried out at each step, then the rounding errors would magnify at the end in the calculations.

As mentioned earlier, there is a whole range of values for a physical quantity with measurement due to uncertainty. Mostly, the measurement values are distributed around the average value in the well-known bell-shaped curve also called normal (Gaussian) distribution in which there is a probability that the measured value lies close to average (mean) than at the extremes of the curve. This can be proved by the standard deviation concept in statistics.

Example 9: Round off 6.2349 ± 0.152.
Solution: Since the uncertainty begins with 1, the number of s.f. would be 2. Also, the quantity and uncertainty should have the same number of decimal places, hence, the answer should be 6.23 ± 0.15.

Example 10: Round off the quantity $t = 5.42 \times 10^6$ s, where its uncertainty $\Delta t = 3 \times 10^4$ s.
Solution: Since the quantity as well as its uncertainty is given in scientific notation, it may be written as $(5.42 \pm 0.03) \times 10^6$.

The representation should have the quantity and uncertainty in the same power of 10.

2.5 Classification of errors

Although uncertainty is more standard way in sciences, yet sometimes scientists talk about errors also, so when quantifying errors, they are classified as

(1) Absolute error and (2) Relative error

(1) Absolute error

The absolute error (E_{abs}) refers to the difference between the actual value x and observed value a, then

$$E_{\text{abs}} = x - a$$

(2) Relative error

Relative error (E_{rel}) is defined as the ratio between absolute error and absolute value of the observation. It is a dimensionless quantity:

$$E_{rel} = \frac{x - a}{a}$$

They are usually expressed as either percent (×100%), fractions, parts per thousand (×10^3) or parts per million (×10^6).

Example 11: Demonstrate the absolute and relative errors in the approximation of numerical constant e (e = 2.71828).

Solution:
$$E_{abs} = |2.71 - e| = 0.00828$$

$$E_{rel} = \frac{|2.71 - e|}{|e|} = 0.003$$

Note: When the absolute errors are indeterminate error, they are followed by " ±," if they are with the determinate, then the original sign follows.

2.6 Problems for practice

(1) Express the quantities in the standard form of uncertainty as $x + \Delta x$.
    (a) $X = 23.0567, \Delta X = 0.03279$
    (b) $m = 0.01362\text{kg}, \Delta m = 2.418 \times 10^{-3}\text{kg}$
    (c) $a = 11.27 \times 10^{32}, \Delta a = 3.6142 \times 10^{30}$
    (d) $Y = 9.11 \times 10^{-33}, \Delta Y = 3.2145 \times 10^{-33}$

(2) Which of the following are more accurate and why?
    (a) π, 3.1392842 or 3.1417
    (b) 2.7182820135423 or 2.718281828 (as an approximation of e).

Chapter 3
Some mathematical functions

3.1 Introduction

A mathematical function is a relationship or an expression between a set of variables (dependent and independent). They are integral in establishing relationships in sciences. A mathematical function may be as simple as a linear function or may be as complex as a polynomial. It may be considered as a device that converts one value to another. A function is often called an argument; especially in physical chemistry, the mathematical functions are used to model physical parameters and draw inferences from the model. A function is often written in the form

$$y = f(x) \tag{3.1}$$

which implies that y (dependent variable) is a function of x (independent variable).

3.2 Classification of functions

Although there are many classes of classification, here, only the elementary classification of functions is mentioned.

3.2.1 Linear function

The linear function is the simplest function of all the mathematical functions. It is represented as

$$y = mx + c \tag{3.2}$$

When this function is plotted then a straight line is obtained, where c is the intercept and m is the slope of the line (Figure 3.1). If intercept is zero (no intercept), then the line passes through the origin ($y = mx$) and if the slope is zero then the line is horizontal, that is, $y = c$.

Slope could be positive or negative. Figure 3.1(a) and (b) represents negative and positive slopes, respectively. Since the function is linearly changing with x, they are also called first-degree polynomial. Slope of any straight line can be found out by taking two random points on the line. For example, if two points (x_1, y_1) and (x_2, y_2) are considered which if written in the mathematical form are

$$y_1 = mx_1 + c \tag{3.3}$$

https://doi.org/10.1515/9783111334448-003

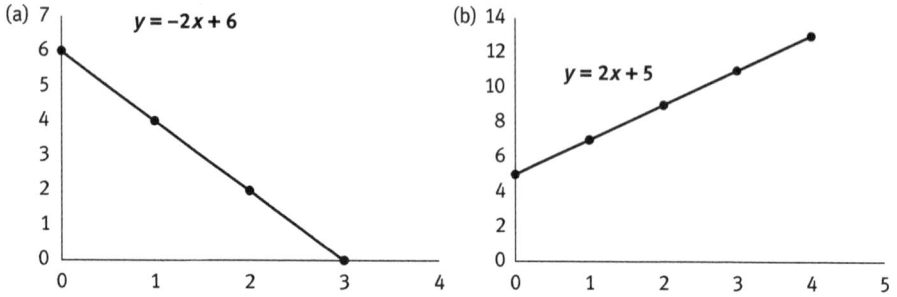

Figure 3.1: Equation of line showing negative slope (a) and positive slope (b).

$$y_2 = mx_2 + c \tag{3.4}$$

then

$$\Delta y = y_2 - y_1 = m(x_2 - x_1) \tag{3.5}$$

$$m = \frac{y_2 - y_1}{x_2 - x_1} \tag{3.6}$$

where m is the slope of the line joining (x_1, y_1) and (x_2, y_2).

3.2.2 Exponential

Mostly the term exponential is used to indicate the fast pace of increasing or decreasing. In the exponential function, the independent variable is given as exponent to some other constant or base (b in this case). It can be represented as

$$y = b^x \tag{3.7}$$

$$y = e^x \tag{3.8}$$

If $x = 0$, $y = 1$; $y = \infty$ if $x \to \infty$ and $y = 0$ if $x \to -\infty$

where x is an independent variable (can range from entire range of real numbers) and y is a dependent variable (on x). Figure 3.2 shows the exponential graph for $y = e^x$ (a) and $y = e^{-x}$ (b).

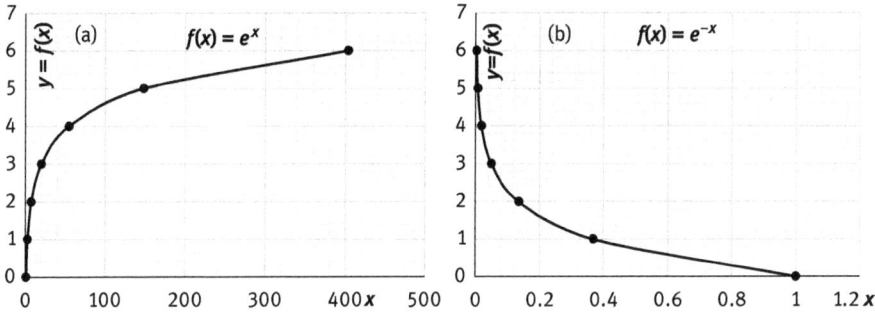

Figure 3.2: Exponential (a) $f(x) = e^x$ and (b) $f(x) = e^{-x}$.

If $b = 2$, then the function is $f(x) = y = 2^x$ (Figure 3.3(a)) and if $b = 0.5$, then the function $f(x) = y = 0.5^x$ (Figure 3.3(b)).

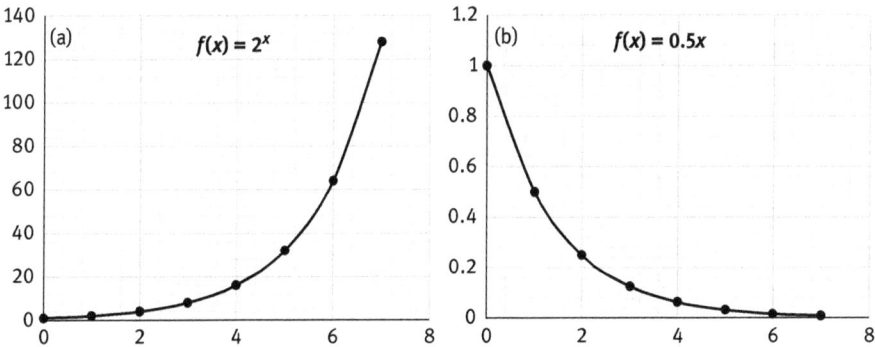

Figure 3.3: Exponential (a) $f(x) = 2^x$ and (b) $f(x) = 0.5^x$.

Consider the example of 10^x, which is used to represent decimal number where x is an integer. By using calculus, it can be found that the natural choice for base b in sciences turns out to be an irrational number as $e = 2.718281828459045$. Hence, it can be written as $f(x) = y = e^x$ or more generally as $y = e^{kx}$. It is also known as *Euler's number* (e), which is defined as the infinite series as

$$e = 1 + \frac{1}{1!} + \frac{1}{2!} + \frac{1}{3!} + \frac{1}{4!} + \cdots \tag{3.9}$$

$$e = 2.718281828459045 \tag{3.10}$$

Some of the examples of equation involving exponential mathematical function include Boltzmann's equation for population (3.11), Arrhenius's equation (3.12) and exponential radioactive decay (3.13) to name a few:

$$N = N_0 \exp(-\Delta E / RT) \tag{3.11}$$

$$k = A \exp(-E_a / RT) \tag{3.12}$$

$$\lambda = \lambda_0 \exp(-kT) \tag{3.13}$$

3.2.3 Logarithmic function

The logarithm is the inverse of exponential. It tells how many times a number should be multiplied to itself to get a particular answer. The base of the log is the main basis of classification. One may write the base as either 10 or e.

(a) Logarithms to the base 10
If the base of log is 10 then they are called common logarithms (Figure 3.4). For example, $10^y = x$, then y is called common log of x denoted by $\log_{10}(x)$. The subscript 10 is often not written to simplify. For example, in chemistry, pH of the equation is

$$pH = -\log_{10}[H^+] \tag{3.14}$$

which can be simply written as

$$pH = -\log[H^+] \tag{3.15}$$

When no base is mentioned, then it is usually understood that base (a) = 10.

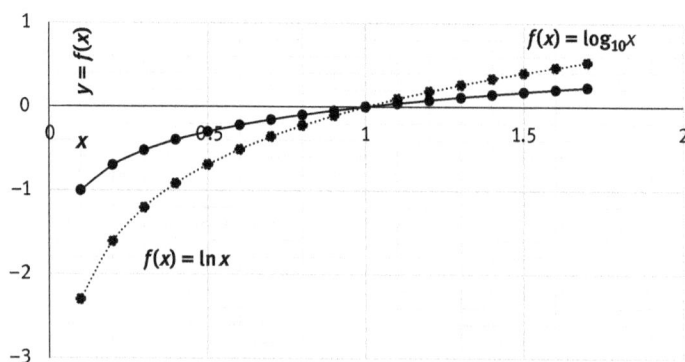

Figure 3.4: $f(x) = \ln x$ (dotted line) and $f(x) = \log_{10}x$ (solid line).

(b) Logarithms to the base e
If the base of the logarithm is e then it is known as *natural logarithms*:

$$\ln(x) = \log_e(x) \tag{3.16}$$

Natural log is the inverse for exponential

$$y = e^x \tag{3.17}$$

Then

$$x = \ln(y) \tag{3.18}$$

$$\ln(e^x) = x \tag{3.19}$$

or,

$$e^{\ln(x)} = x \tag{3.20}$$

For example, $\log_2 16$ implies 2 (base value) should be multiplied itself 4 times to get answer 16. The more general form is

$$\log_a(x) = y \text{ or } x = a^y \tag{3.21}$$

where a is the base of the logarithm. The logarithm is the inverse of exponential, so they both can be cancelled if the base is same.

Example 1: Simplify

$$f(x) = \log_{10}\left(10^{2x+5}\right) \tag{3.22}$$

Solution: Since both exponents and logarithm have the same base 10, they cancel out each other:

$$f(x) = 2x + 5 \tag{3.23}$$

Example 2: Simplify

$$f(x) = \log_3\left(3^x . 9x^{100}\right) \tag{3.24}$$

Solution: The above expression can be simplified as

$$f(x) = \log_3\left(3^x\right) + \log_3 9 + \log_3\left(x^{100}\right) \tag{3.25}$$

$$f(x) = x + 2 + 100\log_3 x \tag{3.26}$$

Example 3: Solve

$$f(x) = a^{\log_a\left(2x^2\right) + 6x} \tag{3.27}$$

Solution: The above function can be written as

$$f(x) = a^{\log_a\left(2x^2\right)} . a^{6x} \tag{3.28}$$

$$f(x) = 2x^2 . a^{6x} \tag{3.29}$$

*If the base and exponent are different, then they both will not cancel each other.

3.2.4 Polynomials

The polynomials can be represented as

$$f(x) = a_0 + a_1 x + a_2 x^2 + \cdots + a_n x^n \tag{3.30}$$

Or in general,

$$f(x) = \sum_{i=0}^{n} a_i x^i \tag{3.31}$$

where a_i is constant and n is a positive integer. If $n = 1$, then it is called a linear function (straight line):

$$f(x) = a_0 + a_1 x \tag{3.32}$$

which is of the same form as

$$y = mx + c \tag{3.33}$$

If $n = 2$, then it is called a quadratic function as

$$f(x) = a_0 + a_1 x + a_2 x^2 \tag{3.34}$$

which can be conveniently written as

$$f(x) = ax^2 + bx + c \tag{3.35}$$

$$f(x) = x^2 + 5x + 6 \tag{3.36}$$

This quadratic equation can be solved for the value of x as $x = \left((-b \pm \sqrt{D})/2a \right)$, where $D = b^2 - 4ac$. If $n = 3$ and $n = 4$, then it is called cubic function and quartic function, respectively.

A cubic equation may be represented graphically (Figure 3.5) and mathematically as

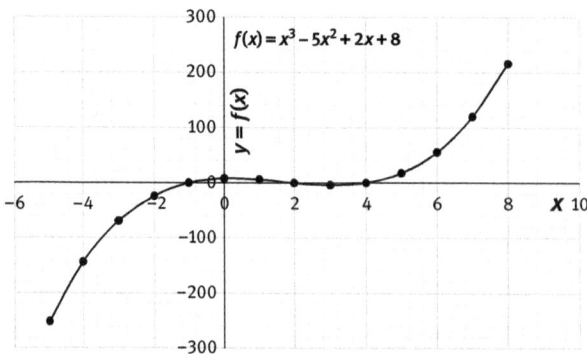

Figure 3.5: $f(x) = x^3 - 5x^2 + 2x + 8$.

$$f(x) = x^3 - 5x^2 + 2x + 8 = (x-2)(x+1)(x-4) \tag{3.37}$$

3.2.5 Power function

A power function may be represented as $y = x^n$, where n is a constant. The linear, quadratic, cubic and quartic functions are examples of power functions. *Reciprocal functions* and *root functions* are sometimes considered as separate class, but they can also be clubbed under the category of power function as $y = \frac{1}{x}$, since $y = x^{-1}$, $y = \sqrt{x}$ since $y = x^{1/2}$, $y = \sqrt[3]{x}$ and so on. They are represented in the graphical form (Figure 3.6).

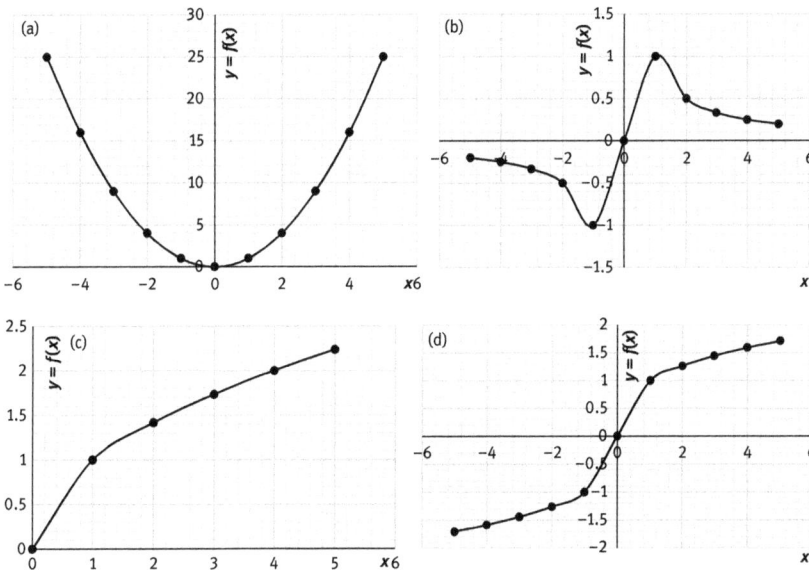

Figure 3.6: Graphical representation of (a) $y = x^2$; (b) $y = 1/x$; (c) $y = \sqrt{x}$; and (d) $y = \sqrt[3]{x}$.

3.2.6 Circle

A circle in a general form can be represented as

$$(x-h)^2 + (y-k)^2 = r^2 \tag{3.38}$$

where (h, k) is the centre and r is the radius of the circle (Figure 3.7).
If the centre is the origin $(0,0)$, then the equation becomes

$$x^2 + y^2 = r^2 \tag{3.39}$$

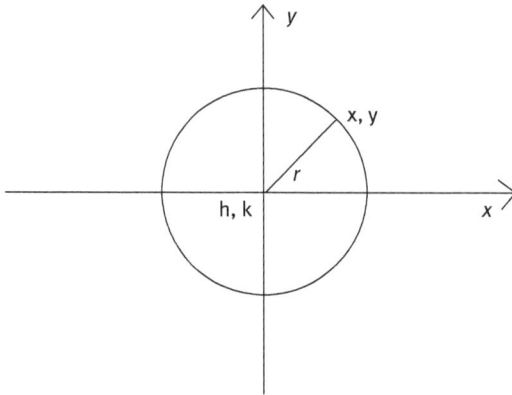

Figure 3.7: Equation of circle $f(x) = x^2 + y^2 = r^2$.

3.2.7 Rational number

A rational number is a number obtained by dividing an integer with another integer (non-zero). If there are two functions $A(x)$ and $B(x)$ as

$$A(x) = a_0 + a_1 x + a_2 x^2 + \cdots + a_n x^n \tag{3.40}$$

$$B(x) = b_0 + b_1 x + b_2 x^2 + \cdots + b_m x^m \tag{3.41}$$

such that $f(x)$ is

$$y = f(x) = \frac{A(x)}{B(x)} = \frac{a_0 + a_1 x + a_2 x^2 + \cdots + a_n x^n}{b_0 + b_1 x + b_2 x^2 + \cdots + b_m x^m} \tag{3.42}$$

For example,

$$\frac{2x-3}{x-5}, \quad \frac{x^2+3x-10}{x+1}, \quad \frac{1}{x}, \quad \frac{x-2}{x^2+2x-3} \tag{3.43}–(3.46}$$

Let us take $f(x) = y = \frac{1}{x}$, as can be seen from the graph (Figure 3.8) when $x \to 0$, then $y \to \infty$.

Similarly, when $x \to 0$ from the minus side, $y \to \infty$. This point of $x = 0$ is called point of singularity, where curve approaches the y-axis. The y-axis, that is, $x = 0$ is called the asymptote to the curve or one may say that the curve approaches the line $x = 0$ asymptotically. All rational functions have at least one point of singularity. It can also be said that as $x \to \infty$ from either side, the curve approaches y-axis but does not cross it. So here $y = 0$ is an asymptote to the curve.

Also, if the degree of polynomial in the numerator is greater than in the denominator, it is called an improper rational function and if it is smaller than in the denominator, it is called proper rational function.

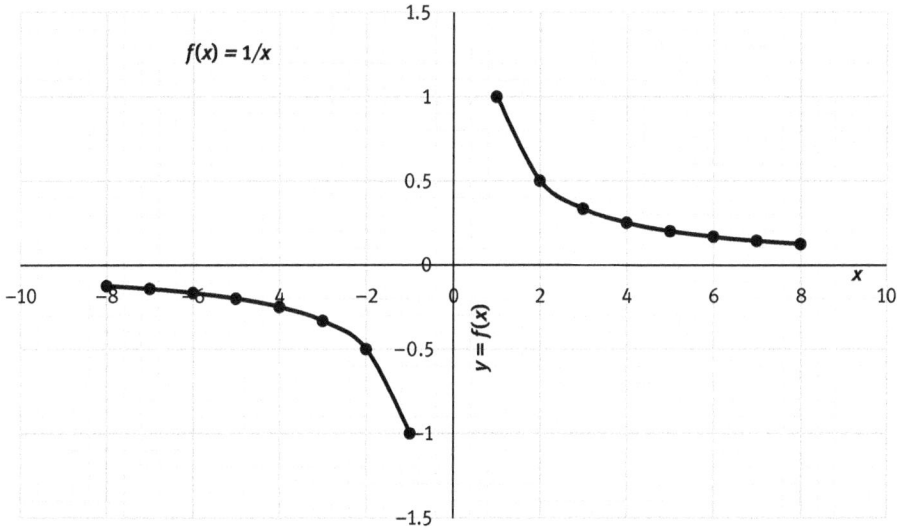

Figure 3.8: Rational number $y = 1/x$.

3.2.8 Partial fractions

Sometimes quotients and quadratic equations in denominator can be represented in more than one form as a collection of several partial fractions. This method of expressing the quotient as partial fractions has application during integration and solving differential equations. A rational function of the form $A(x)/B(x)$ can be factorized into simple partial fractions as

$$\frac{x+5}{(x-2)(x+3)} = \frac{A}{x-2} + \frac{B}{x+3} \tag{3.47}$$

$$\frac{x+5}{(x-2)(x+3)} = \frac{A(x+3)+B(x-2)}{(x-2)(x+3)} \tag{3.48}$$

$$x+5 = A(x+3) + B(x-2) \tag{3.49}$$

At $x = -3$, $B = -2/5$ while at $x = 2$, $A = 7/5$.

3.2.9 Trigonometric functions

There are three principal trigonometric functions namely sine (sin), cosine (cos) and tangent (tan). The other trigonometric functions are secant (sec), cosecant (cosec) and cotangent (cot) are derived functions, where

$$\operatorname{cosec}\theta = \frac{1}{\sin\theta}, \quad \sec\theta = \frac{1}{\cos\theta} \quad \text{and} \quad \cot\theta = \frac{1}{\tan\theta} \qquad (3.50\text{--}3.52)$$

They may be depicted graphically as in Figure 3.9.

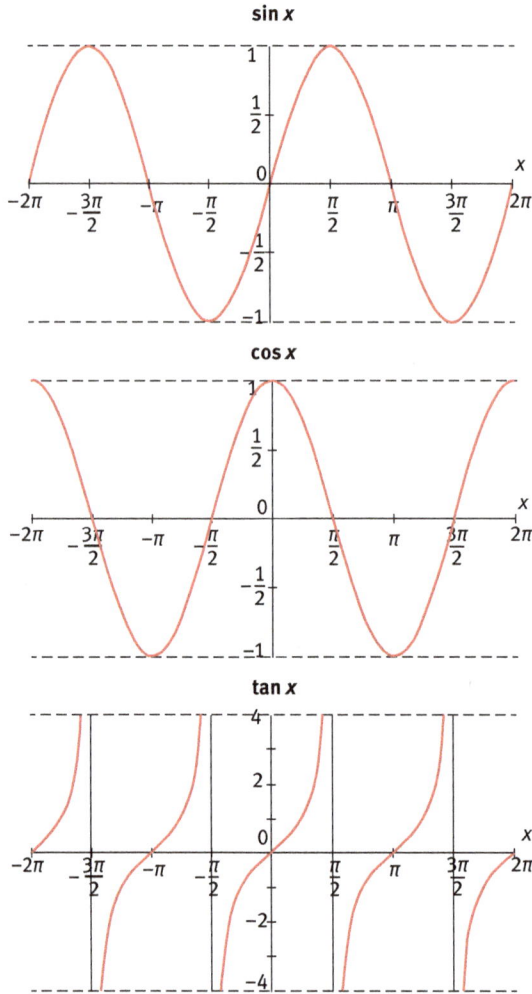

Figure 3.9: Some trigonometric functions (sin, cos and tan functions).

Some important trigonometric identities that are frequently used are

$$\sin^2\theta + \cos^2\theta = 1 \qquad (3.53)$$

$$\operatorname{cosec}^2\theta = 1 + \cot^2\theta \qquad (3.54)$$

$$\sec^2\theta = 1 + \tan^2\theta \tag{3.55}$$

$$\sin a + \sin b = 2\sin\frac{a+b}{2}\cos\frac{a-b}{2} \tag{3.56}$$

$$\sin a - \sin b = 2\cos\frac{a+b}{2}\sin\frac{a-b}{2} \tag{3.57}$$

$$\cos a + \cos b = 2\cos\frac{a+b}{2}\cos\frac{a-b}{2} \tag{3.58}$$

$$\cos a - \cos b = -2\sin\frac{a+b}{2}\sin\frac{a-b}{2} \tag{3.59}$$

Both degree and radian are often used as a unit of angle. Radian is the SI unit of angle (SI symbol rad). In terms of circle, one radian is the angle (θ) subtended by an arc (of length s), which is equal in length to the radius (r) of the circle (Figure 3.10). More precisely, it is equal to the ratio of arc length to the radius of the circle, that is,

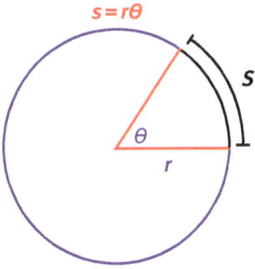

Figure 3.10: Arc of length subtending angle.

$$\theta = \frac{s}{r} \tag{3.60}$$

Since the arc is proportional to the angle, the whole arc of circumference $2\pi r$ will subtend angle of 360°. So 360° is equal to 2π radians. Hence,

$$1\text{rad} = \frac{360^o}{2\pi} \tag{3.61}$$

$$1\text{rad} = \frac{180}{2\pi}\text{ degrees} \tag{3.62}$$

For example, angle 270° is $\frac{3\pi}{2}$ radians.

3.2.9.1 Inverse trigonometric functions

Inverse trigonometric functions are literally the inverse of trigonometric functions, that is, $\sin^{-1}x$, $\cos^{-1}x$, $\tan^{-1}x$, $\csc^{-1}x$, $\sec^{-1}x$ and $\tan^{-1}x$. Also called anti-trigonometric functions, these functions are used to find the angle for the given trigonometric ratio. The prefix "arc" is used to represent the inverse of the function, for example,

$$\sin^{-1}x = \arcsin(x) \tag{3.63}$$

$$\sin^{-1}(1/2) = \pi/6 \tag{3.64}$$

3.2.9.2 Hyperbolic trigonometric functions

Hyperbolic trigonometric functions are analogous to trigonometric functions that are defined in terms of exponential function. There are six hyperbolic functions, namely, $\sinh x$, $\cosh x$, $\tanh x$, $\operatorname{sech} x$, $\operatorname{cosech} x$ and $\coth x$. As the trigonometric functions parametrize a circle, hyperbolic trigonometric functions parametrize hyperbola. They have similar identities as trigonometric functions. The hyperbolic sine and cosine functions are written as

$$\sinh = \frac{e^x - e^{-x}}{2} \tag{3.65}$$

$$\cosh = \frac{e^x + e^{-x}}{2} \tag{3.66}$$

3.2.10 Taylor's series

A Taylor's series is a representation of a function ($f(x)$) around a point (a) in terms of its derivatives that can be written as a sum of infinite terms as

$$f(x) = \sum_{n=0}^{\infty} f^n(a)\frac{(x-a)^n}{n!} = f(a) + f'(a)(x-a) + \frac{1}{2!}f''(a)(x-a)^2 + \frac{1}{3!}f'''(a)(x-a)^3 + \cdots \tag{3.67}$$

where $f'(x)$ and $f'(x)$ are the first and second derivatives of $f(a)$ with respect to x and $h = x-a$. To represent the function $f(x)$ in terms of a polynomial, it should be defined in regions near $x = a$ and when the function $f(x)$ is differentiable as many times, it gives a good approximation of the function in the region near $x = a$.

According to Taylor's theorem, if function fulfils some predefined conditions then it may be expressed as Taylor's series. If $a = 0$, then the Taylor's expansion becomes the Maclaurin series.

3.2.10.1 Maclaurin series

The Maclaurin series is a special case of Taylor's series where the function is centred at zero ($a = 0$):

$$f(x) = f(0) + xf'(0) + x^2 \frac{1}{2!}f''(0) + x^3 \frac{1}{3!}f'''(a) + \cdots \tag{3.68}$$

$$\sin x = x - \frac{x^3}{3!} + \frac{x^5}{5!} - \frac{x^7}{7!} + \cdots = \sum_{n=0}^{\infty} \frac{(-1)^n}{(2n+1)!}x^{2n+1} \tag{3.69}$$

$$\cos x = 1 - \frac{x^2}{2!} + \frac{x^4}{4!} - \frac{x^6}{6!} + \cdots = \sum_{n=0}^{\infty} \frac{(-1)^n}{(2n)!}x^{2n} \tag{3.70}$$

$$e^x = 1 + x + \frac{x^2}{2!} + \frac{x^3}{3!} + \frac{x^4}{4!} + \cdots = \sum_{n=0}^{\infty} \frac{x^n}{n!} \tag{3.71}$$

Algebric functions and transcendental functions

Algebric functions are the functions upon which simple operations of addition, subtraction, division and multiplication are applicable, while the transcendental functions are functions other than algebric functions (transcends means beyond algebra). They cannot be expressed as a solution of a polynomial equation. Transcendental functions involve trignometric functions, inverse functions, hyperbolic functions, logarithmic and exponential functions. These functions can be expressed in algebric terms only in terms of infinite series. For example, e^x or π^x are transcendental functions that can be expressed in terms of infinite series.

3.2.11 Coordinate systems

Coordinate system uses real numbers to describe the position of points or other geometric elements in the space (two dimensional or three dimensional).

(a) Cartesian coordinate system

In Cartesian coordinate system, each point is specified uniquely in a plane by a set of numerical coordinates from the two fixed perpendicular lines. If there are three mutually perpendicular lines in three dimensions, then n number of coordinates can be specified in such n-dimensional space called Euclidean space. The point at which these lines intersect is known as origin (Figure 3.11).

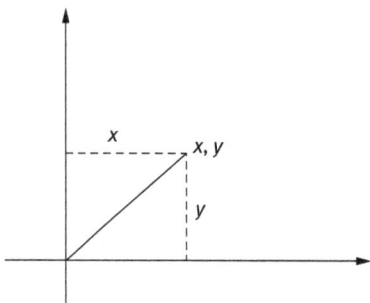

Figure 3.11: Cartesian coordinates.

(b) Polar coordinate system

Polar coordinate system refers to a two-dimensional coordinate system, where a point is specified by the radial distance and angle of the point from reference point and direction, respectively (Figure 3.12):

$$x = r \cos \theta \tag{3.72}$$

$$y = r \sin \theta \tag{3.73}$$

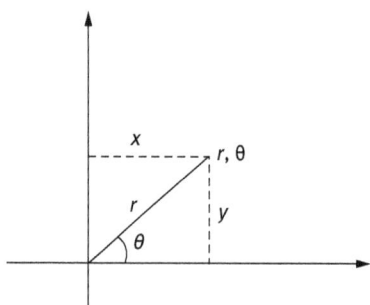

Figure 3.12: Polar coordinates.

(c) Spherical coordinate system

Spherical coordinate system is the coordinate system for three-dimensional space where a point is specified via three numbers, namely, radial distance of a point from the origin, polar angle measured from a zenith direction and azimuthal angle of its orthogonal projection on a reference plane (Figure 3.13):

$$z = r \cos \theta \tag{3.74}$$

$$x = r \sin \theta \cos \varphi \tag{3.75}$$

$$y = r \sin \theta \sin \varphi \tag{3.76}$$

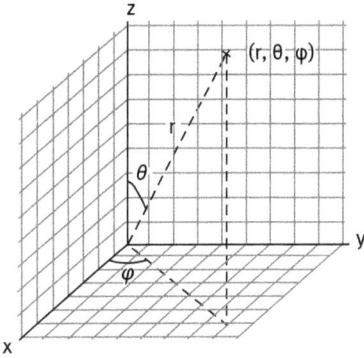

Figure 3.13: Polar coordinates.

3.2.12 Number system

There are many ways to represent a number like whole numbers and real numbers. In context of computers, there is a different sort of classification of numbers which is
(a) Decimal
(b) Binary
(c) Octal
(d) Hexadecimal

They can be interconverted conveniently without any loss to the numeric value.

(a) Decimal number system
Decimal number system uses digits 0 to 9 to represent numbers. Since it uses 10 numbers (0–9), it is called the base 10 number system. Each digit has a value based on its position called place value. For example,

$$895 = 8 \times 10^2 + 9 \times 10^1 + 5 \times 10^0$$
$$= 800 + 90 + 5$$
$$= 895$$

(b) Binary number system
Binary number system is a system of writing numbers in terms of two symbols 0 and 1 (binary means two) and they have a base of 2. This number system is used by electronic, digital and computer devices as a medium of language. All the information or numbers are first converted into binary form and then input into computer's memory. This number system uses positional notation. For example, binary number $(1000101)_2$ is converted into decimal number.

In Table 3.1, the bits (binary numbers) have a positional value and has some weight associated with it which is the power of 2.

Table 3.1: Position and bits for conversion of binary to decimal number.

6	5	4	3	2	1	0	Position
1	0	0	0	1	0	1	Bit

The binary number is converted as

$$=1\times 2^6 + 0\times 2^5 + 0\times 2^4 + 0\times 2^3 + 1\times 2^2 + 0\times 2^1 + 1\times 2^0$$
$$= 64 + 4 + 1$$
$$=(69)_{10}.$$

Likewise a decimal number can also be converted into a binary number system by dividing it by 2 until the quotient is 0. The remainder is noted each time and taken in reverse order.

2	69	
2	34	1
2	17	0
2	8	1
2	4	0
2	2	0
2	1	0
	0	1

$$69 = (1000101)_2$$

Similarly, a decimal fraction is converted into binary fraction by successively multiplying it by 2 until the fraction becomes 0.

	0.875×2
1	$.750 \times 2$
1	$.500 \times 2$
1	$.000$

(c) Octal number system

Octal number system is used to represent digital circuit numbers and is more compact when compared to binary number system. In the octal system, the base is 8. To convert a decimal number into octal number, it is divided by 8 until the remainder is 0 as in binary system.

```
8 | 2065
8 | 258    1
8 | 32     2
  | 4      0
```

$(4021)_8$ is equivalent to $(2065)_{10}$.

Likewise, it can be traced back to decimal system as

$$= 4 \times 8^3 + 0 \times 8^2 + 2 \times 8^1 + 1 \times 8^0$$
$$= 2{,}048 + 16 + 1$$
$$= 2{,}065.$$

(d) Hexadecimal system

Hexadecimal number system is even more compact when compared to octal and binary system. Their base is 16 which signifies 16 single character digits or symbols. The first 10 digits are decimal system digits and the remaining six are denoted by A, B, C, D, E and F as shown (Table 3.2). The conversion of decimal number to octadecimal is similar to the methods followed earlier for binary and octal systems.

For example, **5 DB** represented as **(5 DB)$_{16}$** is the equivalent of decimal number **(720)$_{10}$**.

Table 3.2: Comparison of four number systems.

Decimal	Hexadecimal	Octal	Binary
0	0	0	0
1	1	1	1
2	2	2	10
3	3	3	11
4	4	4	100
5	5	5	101
6	6	6	110
7	7	7	111
8	8	10	1000
9	9	11	1001
10	A	12	1010
11	B	13	1011
12	C	14	1100

Table 3.2 (continued)

Decimal	Hexadecimal	Octal	Binary
13	D	15	1101
14	E	16	1110
15	F	17	1111
16	10	20	10000

3.3 Problems for practice

(1) Evaluate the following:

    (a) $y = \log_2 5 + \log_2 6$ (b) $y = \log_2 \pi$ (c) $y = \log_3(5^{-4})$

(2) Solve the simultaneous equations for x and y

$$2x + 7y = 23 \text{ and } 5x - 8y = -19$$

(3) Constitute an equation for a circle whose centre lies at (2,3) and its radius is 5 cm.

(4) Arrange the van der Waal's equation of state in the cubic equation in V

$$\left(p + \frac{n^2 a}{V^2}\right)(V - nb) = nRT$$

(5) Express c as an explicit function of Λ_m, where Kohlrausch's law for molar conductivity Λ_m of a strong electrolyte at low concentration (c) (where Λ_m^∞ is the molar conductivity at infinite dilution and b is the constant) is given as

$$\Lambda_m = \Lambda_m^\infty - b\sqrt{c}$$

Chapter 4
Descriptive statistics

4.1 Introduction

Descriptive statistics are tools in statistics that summarizes the data into some defined coefficients that could be a representation of the model or results of an experiment. Descriptive statistics are broken down into measures of central tendency and measures of variability (spread). They are used to predict or infer outcomes of a system. It characterizes the system using the data. Sometimes the term "number crunching" is used to describe these features. In science, an enormous amount of data is generated which needs to be organized and interpreted logically and in a meaningful way, in such cases descriptive statistics come to the rescue by quantifying the data into useful coefficients.

A measure of central tendency in statistics indicates the attempt to find the central position of the data set or distribution. They are sometimes also called the measures of central location or summary statistics. These measures depict where the most values of a data set or distribution lies or which is called the central location of distribution. This central value is the representative of the whole distribution. So, by the definition, the mean or average should be representative of measure of central tendency but there are other representatives as well, namely, median and mode, since mean often is not sufficient or even sometimes misleading. Hence, under different conditions and different types of data, different types of measures of central tendency are used. The central value is a single value which gives the idea of approximation of normality. The measures of dispersion of data are quantified by standard deviation and variance, while the minimum and maximum values of the data are explained using skewness and kurtosis.

These descriptors are useful especially in research where they can meaningfully interpret any experiment. Also, they reduce the bulkiness of data and organize them better. One usually prefers to analyse data and reduce the error, but reducing the data and then analysing the error sounds more intelligent way. One cannot always end the error, but can reduce it by improvising the sources of error. But sometimes the error is reduced by interpreting the data in a different way.

Instead of taking 10 observation values for an experiment, mean of those values is preferred (having more significant numbers), thereby reducing the data, and hence also minimizing the error. Hence, these statistical parameters are a way of *data reduction*.

Some of the frequently used descriptors of statistics are described here.

https://doi.org/10.1515/9783111334448-004

4.2 Mean

The mean or the average is the most common measure of central tendency. It is equal to sum of all the values in the data set divided by the number of observations. So, the mean can be written as

$$\bar{x} = \frac{x_1 + x_2 + x_3 + \cdots + x_n}{n} \tag{4.1}$$

or

$$\bar{x} = \frac{\sum_{i=1}^{n} x_n}{n} \tag{4.2}$$

where $\bar{x}$ is the mean, $x_1, x_2, x_3, \ldots, x_n$ are the observations or the data points and n is the total number of data points, and $\sum$ (the Greek letter sigma) refers to the summation. The mean can be calculated for both continuous and discrete data. The above mean is called sample mean ($\bar{x}$) while population mean is represented by μ as

$$\mu = \frac{\sum_{i=1}^{n} X}{n} \tag{4.3}$$

In statistics, sample and population holds different meaning although they both are calculated in the same manner. The mean includes all values in the data set which may or may not be outliers; hence, there is often the possibility of error.

    In Table 4.1, we have considered the weight of 10 persons (A to J) having different weights. The mean comes out to be 60 kg, which is not the correct interpretation of the overall weights since in the raw data the smallest weight is as small as 25 kg, while the highest weight is as large as 94 kg. So in that case the mean is a mediocre value that is not a real reflection of the data, hence, not an appropriate way to depict the typical weight. Therefore, in this case, data is skewed. Therefore, more robust measures of central tendency are used like median. It minimizes the error in prediction of any one value in the data set since it includes all data points in calculation. There are two more types of mean.

Table 4.1: Sample of weight of 10 persons.

Person	A	B	C	D	E	F	G	H	I	J
Weight (kg)	76	63	54	94	31	40	85	73	25	59

4.2.1 Arithmetic mean

Arithmetic mean is simply mean average. It is obtained by adding all the values of data points divided by the number of observations. If we are given raw data of n number of observations, then

$$\bar{X} = \frac{\sum X}{n} \tag{4.4}$$

Sometimes the data is given with the frequency, that is, the number of times a data point is repeated in the data set. The mean in such cases is given by

$$\bar{X} = \frac{\sum fX}{\sum f} \tag{4.5}$$

where f is the frequency, X is the midpoint of the class interval and n is the number of observations. The mean calculated from raw data is different from mean calculated from continuous data with frequency. The mean is not suitable when the data is skewed or when the data set is small since mean resists the fluctuations between different samples.

4.2.2 Weighted mean

Weighted mean is calculated when some data points carry more weightage (importance) than other. For that the term w_i is assigned to the data value x_i. So the weighted mean is given as

$$\text{Weighted mean} = \frac{\sum wx}{\sum w} \tag{4.6}$$

4.2.3 Geometric mean

Geometric mean is defined as the arithmetic mean of data points on a log scale. It is useful for the data that has more utility when interpreted as a product of the numbers and not as sum (unlike arithmetic mean). It is also expressed as the nth root of the product of an observation

$$GM = \sqrt[n]{(x_1)(x_2)...(x_n)} \tag{4.7}$$

$$\log(GM) = \frac{\sum \log x}{n} \tag{4.8}$$

$^{*}GM$ cannot be used if the data points involve any zero or negative value.

4.2.4 Harmonic mean

Harmonic mean is defined as the reciprocal of arithmetic mean of the reciprocals.

$$\text{HM} = \frac{1}{\text{AM}} = \frac{1}{\frac{\sum (1/x)}{n}} = \frac{n}{\sum (1/x)} \tag{4.9}$$

The HM is used when the reciprocal of observations hold more importance than the observation itself. It is particularly useful to estimate the average sample size of groups, each having different sample size.

4.3 Median

Median is the central or middle value in the distribution data when arranged in ascending or descending order of its magnitude. It divides the distribution data in half. In case of odd number of observations, the middle value is the median, while in case of even number of observations, the average of two middle values is taken as median. Unlike mean, median is less vulnerable to outliers and skewed data. It is a preferred measure of central tendency especially when the data is not symmetrical.

Consider the following data of temperature (in °C) for a set of an experiment arranged in descending order as

$$105, 96, 86, 79, 76, 71, 68, 67, 63, 57, 50$$

In such case, the total number of observations is odd, so median is defined as

$$\text{Median} = (N+1)/2\text{th observation} \tag{4.10}$$

which is 71 °C in the above case ($N = 11$).

If there are even number of observations as

$$105, 96, 86, 79, 76, 71, 68, 67, 63, 57$$

then the median would be defined as

$$\text{Median} = \frac{\left(\frac{N}{2}\right)\text{th} + \left(\frac{N+1}{2}\right)\text{th}}{2} \text{ observation} \tag{4.11}$$

that is, the average of two middle observations is taken, where $(N/2)$th and $((N+1)/2)$th are the positions of the observations. So, the median in the above case would be

$$\text{Median} = \frac{76+71}{2}$$

$$\text{Median} = 73.5 \ ^\circ\text{C}$$

4.4 Mode

The mode is the most frequently occurring observation in the data set. The advantage of using mode as measure of central tendency is that it can be used for both numerical and non-numerical (categorical) data. It is mostly used for categorical data. Graphically, a histogram is used to represent a data set where the highest bar represents mode. But sometimes, there may be two values/variables that have the same frequency (they repeat a number of times); in such cases, mode is not useful.

Sometimes, mode lies way beyond the central tendency or region and hence does not give the accurate idea of the data set.

In Figure 4.1, it shows the number of students voting for their favourite subject. One can clearly see that the popular choice of students is Physical Chemistry than Inorganic Chemistry (since most of the students voted for it). The least preferred choice is Nanochemistry.

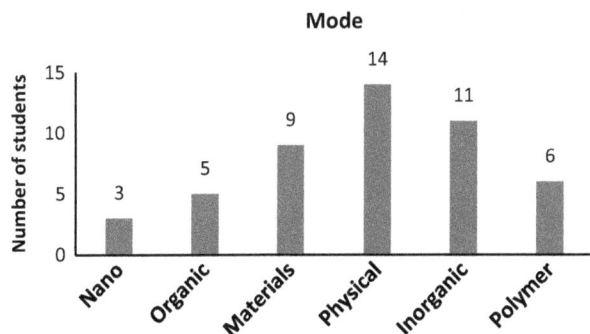

Figure 4.1: Histogram showing mode.

4.5 Standard deviation

When a set of observations are taken for the same measurement then the measured values are grouped around the central value (called mean). Standard deviation describes the deviation of observations from the mean value or how widely the observations are spread on either side of mean. Larger standard deviation implies more widely spread data. Usually, there is a population of N data points where each data point is known for which the standard deviation (σ) is defined as

$$\sigma = \sqrt{\frac{\sum_{i=1}^{n} (x_i - \mu)^2}{N}} \tag{4.12}$$

where N refers to the total number of data points or observations in the population, x_i is the ith observation and μ is the population mean of the N observations. But sometimes it is difficult to access population data; hence, a sample is taken (part or subset of population) which is considered as an approximation of the population parameter. There is a difference between the standard deviation of sample as compared to that of population. For the sample, the standard deviation(s) is defined as

$$s = \sqrt{\frac{\sum_{i=1}^{n}(x_i - \bar{x})^2}{n-1}} \tag{4.13}$$

where $\bar{x}$ refers to the sample mean for n number of observations. As the $n{\to}N$, the sample standard deviation approaches population standard deviation, that is, $s{\to}\sigma$.

4.6 Variance

Variance measures dispersion around mean. It is the squared deviation from the mean. It may be defined by squaring the standard deviation. For a given population having N data points, the variance is

$$\sigma^2 = \frac{\sum_{i=1}^{n}(x_i - \mu)^2}{N} \tag{4.14}$$

There is one more parameter similar to variance called co-variance. Co-variance tells how the two variables vary together. It measures the strength of correlation between two variables where

$$\text{Cov}(x, y) = \frac{\sum_{i=1}^{N}(x_i - \bar{x})(y_i - \bar{y})}{N-1} \tag{4.15}$$

4.7 Coefficient of variation

The coefficient of variation is defined as the ratio of standard deviation to the mean. It represents the extent of variability with respect to the mean of the population or sample. Also known as the relative standard deviation, it is useful in comparing the spread of a distribution. Coefficient of variation often finds its applications in analytical chemistry.

$$CV = \frac{\sigma}{\mu} \tag{4.16}$$

Larger the value of CV, greater is the dispersion of data around the mean. Smaller the CV, more exact is the distribution.

For example, there are two sets of data (A and B) given as follows:

Table 4.2: Analysis of coefficient of variation.

						CV	Std dev.
Set A	1	2	3	4	5	52.7	1.5811
Set B	23	24	25	26	27	6.3	1.5811

As evident from Table 4.2, although both sets have same standard deviation from their mean value, their coefficient of variation (53% and 6%, respectively) is different since their individual means are different.

Example 1: A sample of hematite was analysed for iron percentage in the sample. The measurements were taken repeatedly to ensure true value, which are

42.62, 42.73, 42.75, 42.78, 42.79, 42.83, 42.84, 42.87, 42.92, 42.94

Calculate the mean value, spread, median, average deviation, standard deviation and coefficient of variation.
Solution: The data is arranged, first, into ascending order

42.62, 42.73, 42.75, 42.78, 42.79, 42.83, 42.84, 42.87, 42.92, 42.94

To calculate mean,

$$\text{Mean} = \frac{42.62 + 42.73 + 42.75 + 42.78 + 42.79 + 42.83 + 42.84 + 42.87 + 42.92 + 42.94}{10}$$

$$\text{Mean} = 42.80$$

Spread is the difference between the highest and lowest values in the observations:

$$\text{Spread} = 42.94 - 42.62 = 0.32$$

Since there are even number ($N = 10$) of observations,

$$\text{Median} = \frac{\left(\frac{N}{2}\right)\text{th} + \left(\frac{N+1}{2}\right)\text{th}}{2}$$

$$\text{Median} = \frac{5\text{th} + 6\text{th}}{2}\text{observation}$$

$$\text{Median} = 42.81$$

To calculate average deviation, the results are tabulated (Table 4.3).

Table 4.3: To calculate average and standard deviation.

x_i	$x_i - \bar{x}$	$(x_i - \bar{x})^2$
42.62	−0.18	0.0324
42.73	−0.07	0.0049
42.75	−0.05	0.0025
42.78	−0.02	0.0004
42.79	−0.01	1E-04
42.83	0.03	0.0009
42.84	0.04	0.0016
42.87	0.07	0.0049
42.92	0.12	0.0144
42.94	0.14	0.0196
$\bar{x}_i = 42.80$	$\sum|x_i - \bar{x}| = 0.73$	$\sum(x_i - \bar{x})^2 = 0.0817$

Hence, average deviation $= \dfrac{0.73}{10} = 0.073$

Standard deviation $s = \sqrt{\dfrac{\sum_{i=1}^{n}(x_i - \bar{x})^2}{n-1}}$

$$s = \sqrt{\dfrac{0.0817}{9}} = 0.095$$

$$s^2 = 0.009$$

$$CV = \dfrac{0.095}{42.8} = 0.0021 = 0.21\%$$

Skewness
Skewness indicates the asymmetric distribution. It is given by g as

$$g = \dfrac{\sum_{i=1}^{n}(x_i - \bar{x})^3}{(n-1)s^3} \tag{A.1}$$

If the distribution is Gaussian (symmetrical), then $g = 0$. If g is negative, then left tail of distribution is longer than the right and vice versa.

Kurtosis
Kurtosis represents the pointedness of data's distribution. It is given by k as

$$k = \dfrac{\sum_{i=1}^{n}(x_i - \bar{x})^4}{(n-1)s^4} \tag{A.2}$$

High k implies that most of the standard deviations are caused by extreme deviations from the mean while smaller k implies that deviations are nearer the mean and the distribution is rounded.

4.8 Data analysis

All the descriptors mentioned earlier are the most frequently used parameters for analysis of data. Also, the sampling of data is one of the most important practices. There are a set of tests that helps in identifying the anomaly in the data. It will help to know whether to keep a data value or remove the discordant data. The frequently used tests in data sampling are discussed below.

4.8.1 *Q*-test

Also known as the Dixon's test, the *Q*-test is a simple and widely used statistical tool for finding whether a given data point or result should be retained or rejected. It involves examining the total scattered data and finding the outliers or discordant value. An outlier is a deviant value that may be generated from a different set of sample or distribution.

Keeping an outlier in the data set affects the calculations of mean and standard deviation. Hence, outliers are needed to be removed. For running *Q*-test calculations, the data must be normally distributed (arranged in ascending order), where *Q* is given by

$$Q = \frac{|x_q - x_n|}{w} \tag{4.17}$$

where w is the spread of the observations, that is, the difference between the smallest and highest value in the data, while x_q is the suspected value and x_n is the nearest value to the suspected value when the data is arranged in either ascending or descending order.

This ratio is then compared with critical values Q_c (critical value of Q) corresponding to a particular confidence level (CL = 95% usually). If Q is greater than Q_c, the questionable result can be rejected with the indicated degree of confidence. Table 4.4 contains the Q values at various confidence intervals.

Table 4.4: *Critical values of Q_c.

N	Q_c (CL 90%)	Q_c (CL 95%)	Q_c (CL 99%)
3	0.941	0.970	0.994
4	0.765	0.829	0.926
5	0.642	0.710	0.821
6	0.560	0.625	0.740

Table 4.4 (continued)

N	Q_c (CL 90%)	Q_c (CL 95%)	Q_c (CL 99%)
7	0.507	0.568	0.680
8	0.468	0.526	0.634
9	0.437	0.493	0.598
10	0.412	0.466	0.568

*D.B. Rorabacher, Anal. Chem. 63 (1991) 139.

The Q-test does not give satisfactory results for data set having more than 10 points ($N > 10$). Based on the Q-result, more than one data point could be tested for erratic behaviour but only one value is discarded. Hence, suspected data points (outliers) are legitimately rejected by using this test.

Example 2: The density of aluminium metal was measured by taking replicate measurements which are 2.61, 2.74, 3.4, 2.91, 2.21 g cm^{-3}. On the basis of Q-test, should any of the data point be rejected?

Solution: The first step in detecting an outlier by using Q-test is arranging the data in ascending order as

$$2.21, 2.61, 2.74, 2.91, 3.4$$

The data point 3.4 looks anomalous when compared with the other values so it is the suspected value. Hence, looking at the values, 2.21 is the smallest value, 3.4 being the highest. The closest value to the suspect value is 2.91, which gives

$$Q = \frac{|3.4 - 2.91|}{3.4 - 2.21}$$

$Q = 0.30$. Now, one should look for Q_{crit} or Q_c, if the calculated value exceeds or is well below the Q_c. Since here $N = 5$ and $Q_c = 0.71$ at 95% confidence level (Table 4.3), this data point may be included based on Q-test.

The Q-test cannot be performed if there are more than one identical suspected data points.

Example 3: The analysis of calcite sample yielded CaO percentages of 55.95, 56.00, 56.04, 56.08 and 56.23. The last value appears anomalous; should it be retained or rejected at 95% confidence level?

Solution: Arranging the data in ascending order as

$$55.95, 56.00, 56.04, 56.08, 56.23$$

Here, the difference between the highest (56.23) and lowest values (55.95) is 0.28 which is the spread of the data.

The suspected value and its nearest values are 56.23 and 56.08, respectively, which gives

$$Q = \frac{|56.23 - 56.08|}{0.28} = 0.53$$

For $N = 5$, Q_c at 95% confidence level is 0.71. Since $0.54 < 0.71$, this data point should be retained.

4.8.2 Confidence limit

The experimental mean $\bar{x}$ may or may not be close to true mean μ. There is an entire range under which the true mean lie which is defined by experimental mean and standard deviation. This range is called the confidence interval and its limits are called the confidence limit. The probability that the true value lies within the range is called confidence level, usually given by

$$\text{Confidence limit} = \bar{x} \pm \frac{ts}{\sqrt{N}} \tag{4.18}$$

where t is a statistical factor that depends on the number of degrees of freedom and the confidence level desired and s is the standard deviation. The number of degrees of freedom (v) is one less than the number of measurements (N).

Example 4: A sample was analysed for its wt% in triplicate with the following result 82.5, 82.56 and 82.41 wt% of X. There is 95% confidence that true value lies in a range. What range is that?

Solution: Mean and standard deviation for the given three values are 82.49 and 0.075, respectively. Here $N = 3$, so degrees of freedom $v = N-1 = 2$.

Using eq. (4.18)

$$\text{Confidence limit} = \bar{x} \pm \frac{ts}{\sqrt{N}}$$

$$= 82.49 \pm \frac{4.303 \times 0.075}{\sqrt{3}}$$

$$= 82.49 \pm 0.18$$

Hence, there is 95% confidence that the true value lies between 82.67 and 82.31 in the absence of any determinate error.

4.8.3 *t*-Test

Also called Student's *t*-test, it is often used by analyst to analyse the statistics between the results of two different methods. *t*-test helps to reflect if there is any substantial difference between the two methods or if they measure the same thing. There are three types of *t*-test based on the utility which are explained.

(a) When an accepted value of data is given, there is a set of data that needs comparing. In such cases, *t*-test is given as

$$\mu = \bar{x} \pm \frac{ts}{\sqrt{N}} \tag{4.19}$$

Or, one may write

$$\pm t = (\bar{x} - \mu)\frac{\sqrt{N}}{s} \qquad (4.20)$$

Example 5: A method was analysed for determining the amount of aluminium in a drug using thermogravimetric analysis. Five replicate measurements were made with the same concentration and the mean of the result was found to be 9.6 ppm with the standard deviation of ± 0.7. The literature reference is 10.5 ppm. Is the method statistically correct in reference to the true value at 95% confidence level?
Solution: Using t-test (eq. (4.20))

$$t = (9.6 - 10.5)\frac{\sqrt{5}}{0.7}$$

$$t = 2.9$$

At $N = 5$, the tabulated t value (Table 4.7) is 2.77 at 95% confidence level. Since $t_{calc} > t_{tab}$, there is 95% confidence that the procedure is providing statistically different results from the true value (reference value).

(b) There are one more type of t-tests where the two sets of data compared have *different variances*. Such data sets may have different degrees of freedom as well. They also use the same above formula except for them, $\frac{\sqrt{N}}{s}$ is replaced by $\sqrt{\frac{N_1 N_2}{N_1 + N_2}}/s_p$, where

$$s_p = \sqrt{\frac{\sum(x_{i1} - \bar{x_1})^2 + \sum(x_{i2} - \bar{x_2})^2 + \sum(x_{i3} - \bar{x_3})^2 + \cdots + \sum(x_{ik} - \bar{x_k})^2}{N - k}} \qquad (4.21)$$

Hence,

$$\pm t = \frac{\bar{x_1} - \bar{x_2}}{s_p}\sqrt{\frac{N_1 N_2}{N_1 + N_2}} \qquad (4.22)$$

Example 6: The analysis of iron sample was carried out using two different analytical methods for which replicate measurements were made for each method (A and B) and their data measurements are given as:

Method A	20.10	20.50	18.65	19.25	19.40	19.99
Method B	18.89	19.20	19.00	19.70	19.40	–

Compare the two methods on the basis of their mean values.
Solution: When two methods are to be compared having different mean values, both F-test and t-test can be used. The results are tabulated in Table 4.5.

Table 4.5: *t*-Test for different mean values of two methods.

x_{i1}	$x_{i1}-\overline{x_1}$	$(x_{i1}-\overline{x_1})^2$	x_{i2}	$x_{i2}-\overline{x_2}$	$(x_{i2}-\overline{x_2})^2$
20.10	0.45	0.202	18.89	0.35	0.122
20.50	0.85	0.722	19.20	0.04	0.002
18.65	1.00	1.000	19.00	0.24	0.058
19.25	0.40	0.160	19.70	0.46	0.212
19.40	0.25	0.062	19.40	0.16	0.026
19.99	0.34	0.116			
$\overline{x_1}=19.65\%$		$\sum=2.262$	$\overline{x_2}=19.24\%$		$\sum=0.420$

Using eq. (4.21)

$$S_p = \sqrt{\frac{\sum(x_{i1}-\overline{x_1})^2+\sum(x_{i2}-\overline{x_2})^2}{N_1+N_2-2}}$$

$$S_p = \sqrt{\frac{2.262+0.420}{6+5-2}} = 0.546$$

$$t = \frac{19.65-19.24}{0.546}\sqrt{\frac{6\times5}{6+5}} = 1.23$$

The tabulated t is 2.262 for 9 degrees of freedom (N_1+N_2–2) at 95% confidence limit; hence $t_{calc} < t_{tab}$. Therefore, there is not much difference between two methods of analysis as t_{calc} is well below the t_{tab} value.

---

c) Paired *t*-Test

There are some data sets that need to be compared with other sets of accepted data sets to establish their genuineness. In such cases, each data point is cross validated with the accepted data point for deviation and the t value is given as

$$t = \frac{\overline{D}}{s_d}\sqrt{N} \tag{4.23}$$

where

$$s_d = \sqrt{\frac{\sum(D_i-\overline{D})^2}{N-1}} \tag{4.24}$$

Example 7: Given are the two sets of data taken for a set of samples by two different methods. Determine if there is a considerable difference between two methods.

Sample	A	B	C	D	E	F
Method X	10.2	12.7	8.6	17.5	11.2	11.5
Method Y	10.5	11.9	8.7	16.9	10.9	11.1

Solution: Using eq. (4.23) for paired t-test, the results may be tabulated for various parameters as (Table 4.6):

Table 4.6: Paired t-test results.

Sample	Method X	Method Y	D_i	$D_i - \bar{D}$	$(D_i - \bar{D})^2$
A	10.2	10.5	−0.3	−0.6	0.36
B	12.7	11.9	0.8	0.5	0.25
C	8.6	8.7	−0.1	0.4	0.16
D	17.5	16.9	0.6	0.3	0.09
E	11.2	10.9	0.3	0.0	0.0
F	11.5	11.1	0.4	0.1	0.01
			$\sum D_i = 1.7$		$\sum (D_i - \bar{D})^2 = 0.87$

$$\bar{D} = \frac{1.7}{6} = 0.28$$

$$s_d = \sqrt{\frac{0.87}{6-1}} = 0.42$$

$$t = \frac{0.28}{0.42} \times \sqrt{6} = 1.63$$

The tabulated value of t at 95% confidence limit is 2.571 and since $t_{calc} < t_{tab}$, there is not much considerable difference between two methods.

4.8.4 F-test

It is often desirable to compare the result of one method with those of accepted methods (standard method) to design more robust method of analysis. To determine if one set of results is significantly different from another depends not only on the difference in means but also on the amount of data available and spread. F-test evaluates

Table 4.7: Values of t for $v(N-1)$ degrees of freedom for various confidence levels*.

v	Confidence level			
	90%	95%	99%	99.50%
1	6.314	12.706	63.657	127.32
2	2.92	4.303	9.925	14.089
3	2.353	3.182	5.841	7.453
4	2.132	2.776	4.604	5.598
5	2.015	2.571	4.032	4.773
6	1.943	2.447	3.707	4.317
7	1.895	2.365	3.5	4.029
8	1.86	2.306	3.355	3.832
9	1.833	2.262	3.25	3.69
10	1.812	2.228	3.169	3.581
15	1.753	2.131	2.947	3.252
20	1.725	2.086	2.845	3.153
25	1.708	2.06	2.787	3.078
∞	1.645	1.96	2.576	2.807

* *Analytical Chemistry* 7e by Gary D. Christian et al., John Wiley & Sons.

the difference between spread of results. Hence, this test is designed to indicate whether there is a significant difference between two methods based on their standard deviation. It is used to compare two variances.

F is defined in terms of the variances of the two methods (variance is the square of the standard deviation)

$$F = \frac{s_1^2}{s_2^2}$$

(4.25)

where $s_1 > s_2$.

There are two different degrees of freedom v_1 and v_2, where degrees of freedom are defined as $N-1$ for each case.

If the calculated F value from the equation exceeds a tabulated value at the selected confidence level, then there is a significant difference between the variances of the two methods. To carry out an F-test, a null and alternate hypothesis is created to determine the significance level. Also the degrees of freedom in both numerator and denominator are found out. The calculated value is evaluated for a given confidence limit to either discard the null hypotheses or vice versa. While evaluating F-test value, the greater variance should be kept in the numerator.

The null hypotheses assume that $\sigma_1^2 = \sigma_2^2$, while according to the alternate hypotheses $\sigma_1^2 \neq \sigma_2^2$. If $F_{cal} > F_{crit}$, the null hypotheses is rejected where F_{crit} is the value of F at the required level of confidence or significance. Table 4.8 lists F-values at 95% confidence limit with different degrees of freedom.

Table 4.8: Values of F at the 95% confidence level*.

v2 = 2	v1 = 2	3	4	5	6	7	8	9	10	15	20	30
	19	19.2	19.2	19.3	19.3	19.4	19.4	19.4	19.4	19.4	19.4	19.5
3	9.55	9.28	9.12	9.01	8.94	8.89	8.85	8.81	8.79	8.7	8.66	8.62
4	6.94	6.59	6.39	6.26	6.16	6.09	6.04	6	5.96	5.86	5.8	5.75
5	5.79	5.41	5.19	5.05	4.95	4.88	4.82	4.77	4.74	4.62	4.56	4.5
6	5.14	4.76	4.53	4.39	4.28	4.21	4.15	4.1	4.06	3.94	3.87	3.81
7	4.74	4.35	4.12	3.97	3.87	3.79	3.73	3.68	3.64	3.51	3.44	3.38
8	4.46	4.07	3.84	3.69	3.58	3.5	3.44	3.39	3.35	3.22	3.15	3.08
9	4.26	3.86	3.63	3.48	3.37	3.29	3.23	3.18	3.14	3.01	2.94	2.86
10	4.1	3.71	3.48	3.33	3.22	3.14	3.07	3.02	2.98	2.85	2.77	2.7
15	3.68	3.29	3.06	2.9	2.79	2.71	2.64	2.59	2.54	2.4	2.33	2.25
20	3.49	3.1	2.87	2.71	2.6	2.51	2.45	2.39	2.35	2.2	2.12	2.04
30	3.32	2.92	2.69	2.53	2.42	2.33	2.27	2.21	2.16	2.01	1.93	1.84

* *Analytical Chemistry* 7e by Gary D. Christian et al., John Wiley & Sons.

Example 8: A sample was analysed by two different methods and replicate results were taken from these two methods. On the basis of F-test, compare the given results of two methods (Table 4.9).

Solution: To compare the two methods, F-test is often used by comparing the variances of two methods. The variance of the two methods is calculated as

Table 4.9: F-test for two methods A and B.

Method A	138	131	136	139	131
Method B	134	132	138	133	137

$$s_1^2 = \frac{\sum_{i=1}^n (x_{i1} - \bar{x})^2}{N_1 - 1} = \frac{58}{5-1} = 14.5$$

$$s_2^2 = \frac{\sum_{i=1}^n (x_{i2} - \bar{x})^2}{N_2 - 1} = \frac{26.8}{5-1} = 6.7$$

$$F = \frac{s_1^2}{s_2^2} = 2.164$$

The tabulated result (for $v_1 = 4$ and $v_2 = 4$) is 6.39. Hence, the F-values for methods that are compared are way lower than the tabulated allowed results; hence, the two methods A and B are equally established procedures on the basis of their variances.

4.9 Distribution of errors

Often the events and errors follow some trend in the experiment. The various types of distribution that are seen can be summarized as
(1) Gaussian distribution
(2) Binomial distribution
(3) Poisson distribution

4.9.1 Gaussian distribution

In statistics, the Gaussian distribution (also called the normal distribution) is the continuous distribution whose probability density function is

$$p(x) = \frac{1}{\sigma\sqrt{2\pi}} e^{-(x-\mu)^2/2\sigma^2} \tag{4.26}$$

$$n \to \infty$$

for all values of x, where μ and σ are mean and standard deviation, respectively. It was introduced by Gauss and it governs a variety of phenomenon. The Gaussian distribution in a normalized form can be written as

$$f(x) = \int\limits_{-\infty}^{+\infty} p(x)dx = \frac{1}{\sigma\sqrt{2\pi}} \int\limits_{-\infty}^{+\infty} e^{-(x-\mu)^2/2\sigma^2} dx = 1 \tag{4.27}$$

When $\sigma = 1$, it is called standard distribution. It characterizes the distribution of random variable about its mean value (symmetric about the mean). This distribution is the most common distribution in statistics; hence, if one knows the type of distribution is normal (Gaussian), then distribution of that variable can be characterized by analysing its mean and standard deviation. In Gaussian distribution, variance and mean do not necessarily be equal. It has a symmetrical bell-shaped curve (Figure 4.2).

It describes the continuous data having symmetric distribution. The Gaussian distribution describes the distribution of errors in a sequence of random experiments. These errors as discussed earlier could be systematic or random errors. The Gaussian distribution is symmetrical around its mean and that mean divides the symmetric curve into two halves. The total area under the curve bounded by $x = a$ and $x = b$ is equal to 1. The curve is often standardized to make it simpler and it is standardized when $\mu = 0$ and $\sigma = 1$. Hence, the equation would look like as

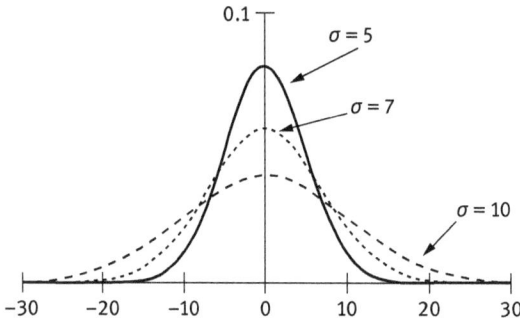

Figure 4.2: Gaussian distribution of errors.

$$f(x) = \frac{1}{\sqrt{2\pi}} \int_a^b e^{-x^2/2} dx \qquad (4.28)$$

4.9.2 Binomial distribution

It describes the distribution of random discrete binary data from a finite sample either success or failure, hence the name binary distribution. According to formula, it shows the probability of getting x events out of n trials. It is most suited when number of outcomes are small usually a success and a failure:

$$P(x) = {}^nC_x p^x q^{n-x} \qquad (4.29)$$

where n is the number of trials, x is the number of successes from n trials, p is the probability for success and q is the probability of failure, $(n-x)$ refers to the number of failures:

$$\sigma = \sqrt{np(1-p)} = \sqrt{npq} \qquad (4.30)$$

where

$$\mu = np \qquad (4.31)$$

If a dice is rolled 30 times, what is the probability of getting 3 on the dice 8 times? Here $N = 30$, $p = 1/6$ and $r = 8$. For large n, binomial distribution approximates to Gaussian distribution.

4.9.3 Poisson's distribution

It is a discrete probability distribution that expresses the probability of a given number of events occurring in a fixed interval of time, distance, area, volume, space and so on. It models random and independent events. It is appropriate for random processes that have rare possibility of occurrence. It is useful to characterize the events with low probabilities of occurrence.

Consider the radioactive decay of nucleus. If the lifetime of nucleus is some years and our measurement takes around 1 min, then probability that the nucleus will decay during measurement is very small; hence, here Poisson distribution works:

$$P(x) = \frac{e^{-\lambda}\lambda^x}{x!} \qquad (4.32)$$

where

$$x = 0, 1, 2, 3 \ldots$$

Above is the probability of observing exactly x events (successes) and λ is the average number of events or successes. It is a limiting form of binomial distribution when n tends to infinity. In Poisson process, mean = variance ($\sigma = \sqrt{\mu}$).

For example, the average number of event in a year is 5 at some place. What is the probability of having 3 events at the same place?

Here, average $\lambda = 5$, $x = 3$:

So

$$P(x = 3) = \frac{e^{-5} 5^3}{3!} = 0.1404$$

Hence, 14% is the probability that event may occur at the same place. According to Poisson distribution, an event may occur at any point of time and one event is independent of the occurrence of another event. It describes the distribution of binary data from an infinite sample. It shows the probability of getting r events from a population (infinite sample).

4.10 Statistical tools using spreadsheets

(1) Average
To find the mean of a sample or population, the function **AVERAGE** is used. The data is first entered into the columns and then the AVERAGE function is used as shown. Its syntax is **AVERAGE(Array1)**

(2) Mode
To find the mode for the given data, **MODE** function is used. Its syntax is **MODE (Array1)** as shown in Figure 4.3.

	A	B
1	*x* values	
2	2.1	
3	2.5	
4	2.3	
5	2.5	
6	=AVERAGE(A1:A5)	=MODE(A1:A5)
7		

Figure 4.3: Mean and mode in spreadsheet.

The different types of **distribution** parameters can also be calculated using spreadsheet. They are explained as follows:

(3) Normal distribution
The syntax is =NORMDIST(Number, Arithmetic mean, Standard deviation, returns the Normal Probability distribution function) (Figure 4.4).

	A	B
1	Number	40
2	Mean	45
3	Std dev	47.609
4		=NORMDIST(B1,B2,B3,FALSE)
5		
6		
7		

Figure 4.4: Normal distribution in spreadsheet.

This gives answer **0.008333** in accordance with the normal distribution function. Similarly binomial distribution and Poisson's distribution can also be used. The syntax is =BINOMDIST(number's, trials, probability's, returns the cumulative distribution function) = POISSON(number, arithmetic mean, returns the cumulative distribution function).

(4) *F*-test
F-test in EXCEL can be carried out by using data analysis tab→*F*-test two-sample for variances (Figure 4.5). Select the data arrays and output cell, where the result will be published.

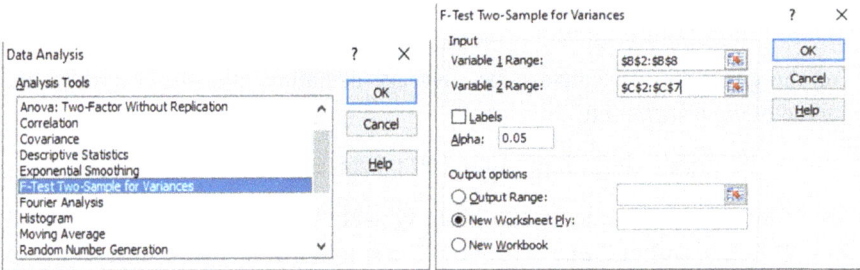

Figure 4.5: Data analysis for *F*-test.

The result obtained is displayed in Figure 4.6. In the picture, df refers to the degrees of freedom which is $N-1$, while F is the value of F-test.

F-Test Two-Sample for Variances

	Variable 1	Variable 2
Mean	10.41666667	10.3875
Variance	0.333666667	0.932678571
Observations	6	8
df	5	7
F	0.357750973	
P(F<=f) one-tail	0.137760942	
F Critical one-tail	0.205091533	

Figure 4.6: *F*-test results in a spreadsheet.

(5) *t*-Test

The t-test mentioned earlier can be performed in spreadsheets for three different options as shown, which complies with three types of test we discussed under t-test (Figure 4.7).

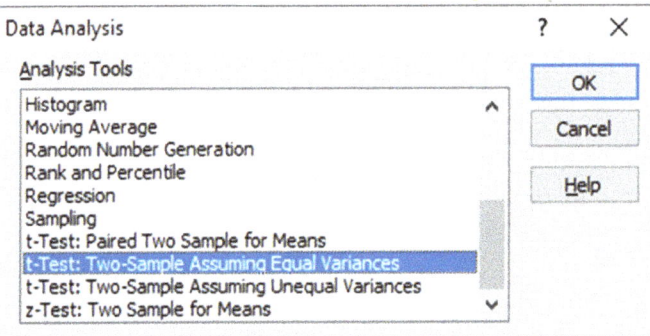

Figure 4.7: *t*-Test in spreadsheets.

4.11 Problems for practice

1. For the given data, calculate mean, average deviation, standard deviation and spread of the distribution

   7.146, 7.098, 6.942, 7.256, 6.593

2. The following measurements of a given variable have been obtained: 23.2, 24.5, 23.8, 23.2, 23.9, 23.5 and 24.0. Apply the Q-test to see if the data point 23.9 is an outlier at 95% confidence level.

3. Calculate the confidence limit for a data set for which mean is 56.06, standard deviation of 0.2% and t-value for 90% confidence limit is 1.833.

4. The analysis of a sample was carried out repeatedly for its % composition in a sample and obtained 2.98, 3.16, 3.02, 2.99 and 3.07. If the true value was 3.03, is the method statistically correct in reference to the true value at 90% confidence level?

5. The following replicate measurements were made on a sample using flame emission spectrometry (FES) and a new colorimetric method and reported. Is there a significant difference in the precision of the two methods?

FES (mg dL^{-1})	Colorimetric (mg dL^{-1})
10.9	9.2
10.1	10.5
10.6	9.7
11.2	11.5
9.7	11.6
10	9.3
	10.1
	11.2
Mean = 10.4	Mean = 10.4

Chapter 5
Numerical curve fitting

5.1 Introduction

Curve fitting refers to a statistical technique constructing a curve or any mathematical function graphically that best describes a series of data points. It establishes the relationship between dependent and independent variables as a function of each other for a given set of data points in the form of mathematical function or mathematical equation. Curve fitting is broadly used for modelling various phenomena. But before understanding these data curve fitting technique, one needs to understand two most important terms which are correlation and regression.

In statistics, correlation and regression are often used to measure the strength of association or establish any relationship between the variables. Before the data is fitted into a mathematical function, a scatter plot between the two variable is drawn. From the scatter plot, it can be inferred whether there exists any relationship between the two variables, also the kind of mathematical relationship between the variables.

5.2 Correlation

It is a statistical measure that establishes the relationship (mostly linear) between two variables hence the term correlation (co + relation). It can be either positive or negative. When two variables involved are moving in the same direction (increase in one also lead to increase in another), then it is called positive correlation and vice versa. Here both variables could be same or different. Correlation is a measure of association between two variables. Here the two variables that are to be correlated can be two random variables. They may not be interdependent on each other. A correlation coefficient of zero indicates no relationship at all, while a positive value indicates that both the dependent (Y) and independent (X) variables move in the same direction (e.g., $r = 0.9$) and a negative coefficient implies dependent and independent variable move in opposite direction (e.g., -0.5). The correlation value of -1 or $+1$ indicates a perfectly linear relationship. If correlation coefficient is zero, then there is no relationship at all. The square of correlation coefficient is the coefficient of determination from regression analysis. There can be many types of correlation relationships but here only two types of correlation are discussed.

https://doi.org/10.1515/9783111334448-005

5.2.1 Pearson correlation coefficient

It evaluates the linear relationship between two continuous variables as the change in one variable produces a proportionate change in another variable. In statistics, the general formula for correlation is given as

$$r = \frac{n\left(\sum_{i=1}^{n} xy\right) - \left(\sum_{i=1}^{n} x\right)\left(\sum_{i=1}^{n} y\right)}{\sqrt{\left(n\sum_{i=1}^{n} x^2 - \left(\sum_{i=1}^{n} x\right)^2\right)\left(n\sum_{i=1}^{n} y^2 - \left(\sum_{i=1}^{n} y\right)^2\right)}} \tag{5.1}$$

Usually a scatter plot is drawn to check for linearity. For evaluating the Pearson coefficient, the variables should be normally distributed apart from having a linear relationship between them. It is a widely used statistical parameter when the data is linear.

Essentially there are two types of correlation coefficient, namely, population correlation coefficient (ρ_{xy}) and sample correlation coefficient (r_{xy}) which can be represented as

$$r_{xy} = \frac{SS_{xy}}{SS_{xx}SS_{yy}} \tag{5.2}$$

$$\rho_{xy} = \frac{\sigma_{xy}}{\sigma_{xx}\sigma_{yy}} \tag{5.3}$$

where (s_n, s_y) and s_{xy} refer to sample standard deviation and co-variance while (σ_{xx}, σ_{yy}) and σ_{xy} are population standard deviation and co-variance.

5.2.2 Spearman's rank correlation coefficient

The Spearman rank correlation coefficient evaluates monotonic relationship between ordinal variables (monotonic relationship implies that there is not a fixed change in one variable with another, although they change together). In Spearman-type correlation, the variation of one variable with the other is not consistent or is always proportionate. Alike Pearson's correlation coefficient, it can also be positive or negative. A correlation value of −1 or +1 indicates linear relationship whether its Pearson or Spearman. It is denoted by ρ as

$$\rho = 1 - \frac{6\sum_{i=1}^{n} d_i^2}{n(n^2 - 1)} \tag{5.4}$$

It is a nonparametric test which is carried out when there is ordinal data (data is arranged in order of ranking or preference) or there is monotonic relationship between data instead of any linear data.

Pearson's coefficient gives the strength and direction of linear relationship between the variables while Spearman's coefficient gives the same for monotonic relationship.

5.3 Regression

Regression predicts the value of dependent (y) variable on the basis of known independent (x) variable value using the mathematical relationship between them. The value of y depends upon the value of x; hence, y is a dependent variable while x is independent. If they are related as

$$y = a + bx \qquad (5.5)$$

then a and b are regression parameters and b is the regression coefficient. It estimates the value of random variable on the basis of fixed variable value. It also involves establishing mathematical relationship between two variables (dependent and independent). A model or relationship is constructed and is then tested for its goodness. If the goodness is good enough, then this model can be used to predict the values of dependent variable for the given independent variable. When a plot of expected value of the dependent variables and independent variable is plotted, the line obtained is called regression line (Figure 5.1). This is called least square curve (LSC) fitting since it fits the given data keeping the square of the error minimum. The mathematical relationship for regression (R^2 or r^2) is given as

$$R^2 = 1 - \frac{ss_{res}}{ss_{total}} \qquad (5.6)$$

It is also known as coefficient of determination where ss_{res} refers to the residual sum of squares (difference between expected and calculated) while ss_{total} refers to the sum of squares (proportional to the variance) where y_i and $f(x_i)$ are the expected and calculated functions:

$$ss_{res} = e_i^2 = (y_i - f(x_i))^2 \qquad (5.7)$$

and

$$ss_{total} = (y_i - \bar{y})^2 \qquad (5.8)$$

The t-test and F-tests are often used to test the validity of regression models.

Here R^2 implies how well the curve fits the data points, also what we know as goodness of fit. Its value ranges from 0 to 1. A value close to 1 implies a good model hence the curve (mathematical function) fits well in data points and vice versa. As the value of R^2 drifts away from 1, the quality of the model falls. Hence a curve is said to have a goodness of fit when its R^2 value approaches 1. The more it recedes away from 1, less is the co-variance between two variables.

5.4 Interpolation versus curve fitting

Interpolation is the technique to find the value of variables in between the existing data points. In interpolation, a curve is constructed through the given data points assuming the data points to be accurate. Interpolation aims to find a function that approximates the given data and determine new data points in between. Curve fitting is applied to scattered data that might not be accurate. Hence a smooth curve is needed in curve fitting that approximates the given data point that might or might not pass through all the data points while in case of interpolation the function passes through original data points. This can be illustrated in Figure 5.1.

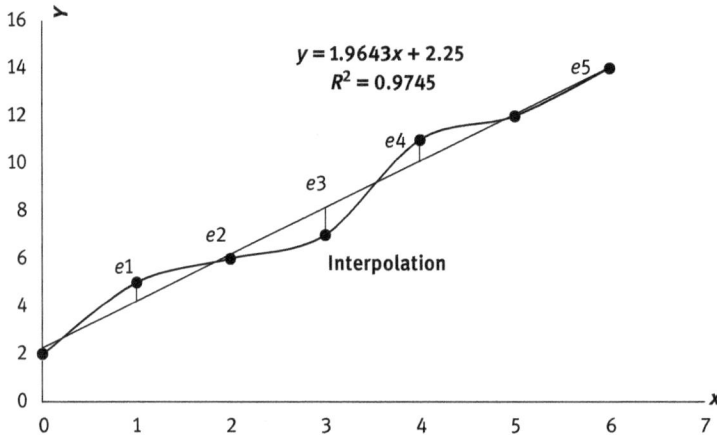

Figure 5.1: Interpolation (curve) and curve fitting (straight line).

The interpolation of points is nothing but connecting the data points together while the other straight line is the curve fitting where a line or a curve has been drawn which best suits the data points that passes through the maximum number of points and depicts some mathematical function (shown by the equation).

5.5 Least square curve fitting

The method of LSC fitting implies constructing a line or a curve that passes through the maximum number of data points and a mathematical relationship which best describes the data points. There are inevitable errors in every experiment or sets of observations. The best we can do is to minimize the error by using LSC. It is applicable for a single curve that represents the general trend of data.

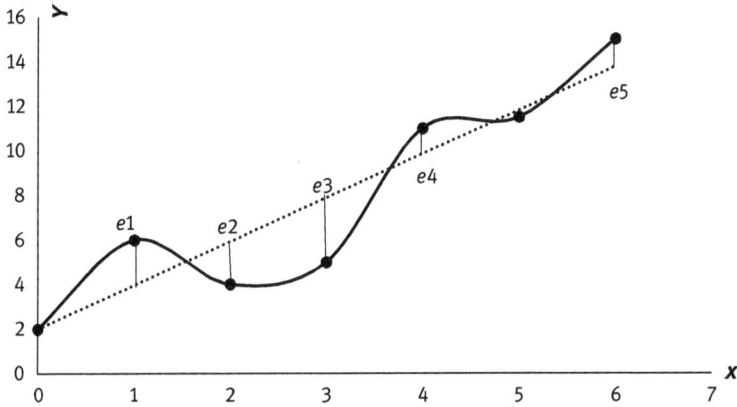

Figure 5.2: Least square curve fitting by minimizing errors.

(1) As can be seen from Figure 5.2, the solid dots represent the actual data while the dotted line is the best possible curve that can be drawn to fit the given experimental data minimizing the error. Here error is the distance between the data point and the best fit line. For a best fit curve, the error should be minimized (also called vertical and perpendicular deviations). This method is also called the LSC method because it aims to minimize the sum of error (also called residual). In the figure, the data is fit into a linear relationship, but regression can also be applied to other functions as well including parabola, polynomial of nth order, exponential and logarithmic function to name a few.

The coefficients a and b in the earlier equations are the values that can be found which best fits the given data into the mathematical model. LSC finds a curve or a mathematical equation that aptly describes the behaviour of the data. Consider the function $f(x)$ written as y

$$y = f(x) \tag{5.9}$$

so if there is a point x_1, then the expected value of y is $f(x_1)$ rather it is found to be y_1. So one can define the error in the value of y (also called residual) as

$$e_1 = y_1 - f(x_1) \tag{5.10}$$

Likewise the error at other points from x_1 to x_n can be written as

$$e_2 = y_2 - f(x_2) \tag{5.11}$$

$$e_3 = y_3 - f(x_3) \tag{5.12}$$

$$e_n = y_n - f(x_n) \tag{5.13}$$

The best fitting curve is the one which has minimum sum of square of offsets (residuals). These residuals can be positive or negative, hence to minimize the error at any point x_i, the square of e_i's is taken as

$$E = \sum_{i=1}^{n} e_i^2 \tag{5.14}$$

This E should be minimized in such a way that the sum of squares of residuals is minimum or least (hence the name least square curve method) by minimizing the square of error. The best line must have minimum error between line and data points.

5.5.1 Linear regression

A straight line is the simplest way of fitting the data by using LSC method using the simple equation $y = a + bx$, where a and b are constants. The number of constants equals the number of equations that needs solving. These equations are called normal equations. Starting from the function

$$y = a + bx \tag{5.5}$$

e_i is the error or residual between the expected value and measured value.

$$e_i = y_{\text{measured}} - y_{\text{expected}} = y_i - (a + bx_i) \tag{5.15}$$

Here some differences may be positive and some may be negative. Hence the error is squared which is minimized

$$\min S_r = \min e_i^2 = \min \left[y_i - (a + bx_i) \right]^2 \tag{5.16}$$

To minimize the error, differentiate the square of the error with respect to a, b and c and put it equal to zero. In regression, one minimizes the sum of squared residuals (S_r).

$$\frac{dS_r}{da} = \frac{dS_r}{db} = 0 \tag{5.17}$$

$$\frac{dS_r}{da} = \sum_{i=1}^{n} [-2(y_i - (a + bx_i))] = 0 \tag{5.18}$$

$$\frac{dS_r}{db} = \sum_{i=1}^{n} [-2x(y_i - (a + bx_i))] = 0 \tag{5.19}$$

which gives

$$na + \sum_{i=1}^{n} bx_i = \sum_{i=1}^{n} y_i \tag{5.20}$$

$$\sum_{i=1}^{n} ax_i + \sum_{i=1}^{n} bx_i^2 = \sum_{i=1}^{n} x_i y_i \tag{5.21}$$

The earlier equation is also called normal equation. These simultaneous equations are often solved by using them in matrix form as

$$\begin{bmatrix} n & \sum_{i=1}^{n} x_i \\ \sum_{i=1}^{n} x_i & \sum_{i=1}^{n} x_i^2 \end{bmatrix} \begin{bmatrix} a \\ b \end{bmatrix} = \begin{bmatrix} \sum_{i=1}^{n} y_i \\ \sum_{i=1}^{n} x_i y_i \end{bmatrix} \tag{5.22}$$

$$\begin{bmatrix} a \\ b \end{bmatrix} = \begin{bmatrix} \sum_{i=1}^{n} y_i \\ \sum_{i=1}^{n} x_i y_i \end{bmatrix} \begin{bmatrix} n & \sum_{i=1}^{n} x_i \\ \sum_{i=1}^{n} x_i & \sum_{i=1}^{n} x_i^2 \end{bmatrix}^{-1} \tag{5.23}$$

Solving eq. (5.23) for a,

$$a = \frac{\sum_{i=1}^{n} x_i^2 \sum_{i=1}^{n} y_i - \sum_{i=1}^{n} x_i \sum_{i=1}^{n} x_i y_i}{n \sum_{i=1}^{n} x_i^2 - \left(\sum_{i=1}^{n} x_i\right)^2} \tag{5.24}$$

$$a = \frac{\bar{y} \sum_{i=1}^{n} x_i^2 - \bar{x} \sum_{i=1}^{n} x_i y_i}{\sum_{i=1}^{n} x_i^2 - n(\bar{x})^2} \tag{5.25}$$

where

$$\bar{x} = \frac{1}{n} \sum_{i=1}^{n} x_i \tag{5.26}$$

and

$$\bar{y} = \frac{1}{n} \sum_{i=1}^{n} y_i \tag{5.27}$$

Solving eq. (5.23) for b,

$$b = \frac{n \sum_{i=1}^{n} x_i y_i - \sum_{i=1}^{n} x_i \sum_{i=1}^{n} y_i}{n \sum_{i=1}^{n} x_i^2 - \left(\sum_{i=1}^{n} x_i\right)^2} \tag{5.28}$$

$$b = \frac{\sum_{i=1}^{n} x_i y_i - n \bar{x} \bar{y}}{\sum_{i=1}^{n} x_i^2 - n(\bar{x})^2} \tag{5.29}$$

Also one may write b as

$$b = \frac{ss_{xy}}{ss_{xx}} \tag{5.30}$$

These expressions can be rewritten in the form of sum of squares. Sum of squares or sum of square deviations is a measure of variation of data set. They describe the difference of the predicted and mean of the variable:

$$SS_{xx} = \sum_{i=1}^{n} (x_i - \bar{x})^2 \tag{5.31}$$

$$SS_{xx} = \left(\sum_{i=1}^{n} x_i^2 \right) - n\bar{x}^2 \tag{5.32}$$

$$SS_{xx} = \sum_{i=1}^{n} x_i^2 - \frac{\left(\sum_{i=1}^{n} x_i \right)^2}{n} \tag{5.33}$$

$$SS_{yy} = \sum_{i=1}^{n} (y_i - \bar{y})^2 \tag{5.34}$$

$$SS_{yy} = \left(\sum_{i=1}^{n} y_i^2 \right) - n\bar{y}^2 \tag{5.35}$$

$$SS_{xy} = \sum_{i=1}^{n} (x_i - \bar{x})(y_i - \bar{y}) \tag{5.36}$$

$$SS_{xy} = \sum_{i=1}^{n} x_i y_i - \frac{\sum_{i=1}^{n} x_i \sum_{i=1}^{n} y_i}{n} \tag{5.37}$$

$$SS_{xy} = n \sum_{i=1}^{n} x_i y_i - \sum_{i=1}^{n} x_i \sum_{i=1}^{n} y_i \tag{5.38}$$

The variance is square of standard deviation, which can also be written as

$$\sigma_x^2 = \frac{SS_{xx}}{n} \tag{5.39}$$

$$\sigma_y^2 = \frac{SS_{yy}}{n} \tag{5.40}$$

$$cov(x, y) = \frac{SS_{xy}}{n} \tag{5.41}$$

where σ_x^2 and σ_y^2 are the variances, and $cov(x,y)$ is the co-variance.

As correlation coefficient is the co-variance divided by the standard deviations of the two variables, so the fit is parametrized into correlation coefficient as

$$r^2 = \frac{SS_{xy}^2}{SS_{xx}SS_{yy}} = \frac{cov(x, y)}{SS_{xx}SS_{yy}} \tag{5.42}$$

The sum of squares reflects the variation in y by variation in x.

The coefficient of determination tells how much y variables can be explained by x values. If for example, $R^2 = 0.86$, then it implies that 86% of the dependent variables are explained by the independent variables.

Example 1: Using LSC fitting, find an equation of the straight line $(y = a + bx)$ with the given data

x	y
0	1
1	10
2	15
3	30
4	35

Solution: To find the equation for the given data, the result for constructing normal equations are tabulated in Table 5.1.

Table 5.1: Curve fitting for a straight line.

x	y	x^2	xy
0	1	0	0
1	10	1	10
2	15	4	30
3	30	9	90
4	35	16	140
$\Sigma x = 10$	$\Sigma y = 91$	$\Sigma x^2 = 30$	$\Sigma xy = 270$

Here $n = 5$, using the normal equations as

$$5a + 10b = 91 \tag{5.43}$$
$$10a + 30b = 270 \tag{5.44}$$

which gives $a = 3/5$ and $b = 44/5$.
Therefore the equation would be

$$y = 0.6 + 8.8x \tag{5.45}$$

In Figure 5.3, solid line represents the interpolation of given data points while the dotted line shows the best fit line obtained by LSC fitting.

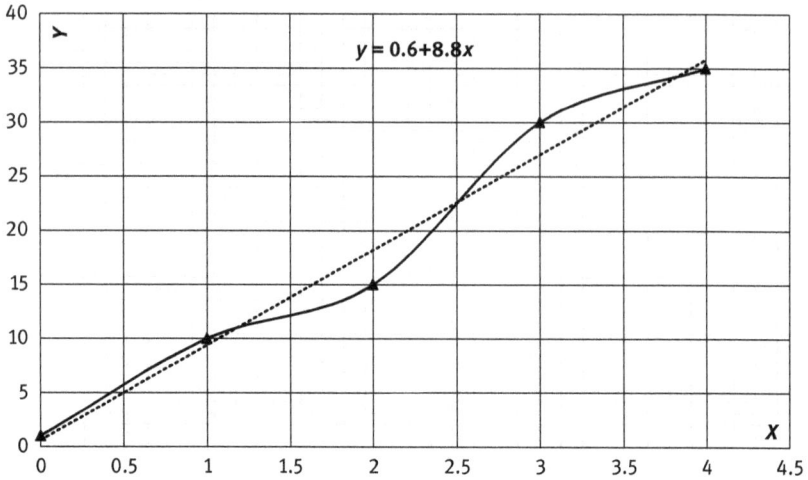

Figure 5.3: X versus Y.

For nonlinear least square fitting to a number of unknown parameters, linear LS fitting may be applied iteratively to a linearized form of the function until convergence is achieved. So if the given data is to be fit into any nonlinear function like

$$y = ax^b \text{ or } y = ae^{bx} \tag{5.46}$$

then the function is first linearized as in the earlier example by taking logarithm on both sides as

$$\log_{10}y = \log_{10}a + b\log_{10}x \tag{5.47}$$

Now the function looks linearized in the form

$$Y = A + bX \tag{5.48}$$

where $Y = \log_{10}y$, $A = \log_{10}a$ and $\log_{10}x = X$

Similarly

$$y = ae^{bx} \tag{5.49}$$

$$\log_{10}y = \log a + bx\log_{10}e \tag{5.50}$$

$$Y = A + BX \tag{5.51}$$

where $\log_{10}y = Y$, $\log_{10}a = A$ and $b\log_{10}e = B$

Example 2: Using the LSC method, fit the following data into the form $y = ax^b$.

x	y
1	3.614
2	8.613
3	14.315
4	20.529
5	27.151
6	34.121

Solution: The earlier data ($n = 6$) is tabulated in Table 5.2 and graphically depicted in Figure 5.4.

Table 5.2: Curve fitting for $y = ax^b$.

x	y	Y	X	XY	X^2
1	3.614	0.558	0.000	0.000	0.000
2	8.613	0.935	0.301	0.282	0.091
3	14.315	1.156	0.477	0.551	0.228
4	20.529	1.312	0.602	0.790	0.362
5	27.151	1.434	0.699	1.002	0.489
6	34.121	1.533	0.778	1.193	0.606
		$\Sigma Y = 6.928$	$\Sigma X = 2.857$	$\Sigma XY = 3.818$	$\Sigma X^2 = 1.775$

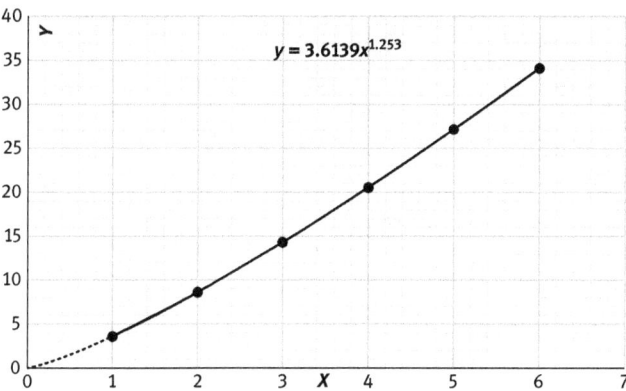

$$y = 3.6139x^{1.253}$$

Figure 5.4: X versus Y.

this gives

$$6A + 2.857B = 6.928 \tag{5.52}$$

$$2.857A + 1.775B = 3.818 \tag{5.53}$$

that gives

$$A = 0.558$$
$$B = 1.253 = b$$
$$a = \text{antilog } A = 3.614$$

Hence

$$y = 3.614x^{1.253} \tag{5.54}$$

Example 3: The vapour pressure of a hydrocarbon varies with temperature as

t^0 (°C)	50	60	70	80
p (kPa)	11.7	19.5	31.7	50.134

Using the linear regression for a straight line

$$\ln(p/\text{kPa}) = -\frac{\Delta_{vap}H_m}{R}\frac{1}{T} + I \tag{5.55}$$

Determine the value of change in enthalpy, that is, $\Delta_{vap}H_m$.

Solution: Since in the data, temperature is given in °C, convert it to K, $T = (t\,°C + 273)K$. Using linear relation $Y = A + BX$, the results are tabulated (Table 5.3), where I represents the intercept a.

Table 5.3: Least square curve fitting to find enthalpy.

$t(^0C)=x$	T(K)	p(kPa)=y	1/T=X	ln p=Y	X^2	XY
50	323	11.7	0.003095975	2.459588842	9.58506E-06	0.007614826
60	333	19.5	0.003003003	2.970414466	9.01803E-06	0.008920164
70	343	31.7	0.002915452	3.456316681	8.49986E-06	0.010076725
80	353	50.134	0.002832861	3.914699421	8.0251E-06	0.0110898
			ΣX_i=0.011847291	ΣY_i=12.80101941	ΣX^2_i=3.51281E-05	0.037701515

Using the normal equations for linear regression ($n = 4$)

$$4a + 0.011847291b = 12.80485006 \qquad (5.56)$$

$$0.011847291a + 0.000035128b = 0.037711994 \qquad (5.57)$$

which on solving for a and b gives, $a = 19.585$ and $b = -5532$.
 Putting back $b = -\frac{\Delta_{vap}H_m}{R}$ which gives $\Delta_{vap}H_m = 45.993$ kJ mol^{-1}, while $I = 3.1831208$.

Example 4: The following data were obtained for decomposition of di-tertiarybutyl peroxide at constant volume:

t (min)	0	3	6	9	12	15	18	21
p (torr)	169.3	189.2	207.1	224.4	240.2	256.0	265.7	282.6

Using the LSC fitting technique for the equation $y = a + bx$ where $y = \log(p_t/\text{Torr})$ with $p_t = \frac{3p_0 - p}{2}$ and $x = t$ (min). Determine the rate constant, if $b = -k/2.303$.
Solution: For using the normal equations for the linear regression these results are as follows (Table 5.4).

Table 5.4: Least square curve fitting for kinetics data.

$t = x$	p	p_t	Y	xy	x^2
0	169.3	169.3	2.228656958	0	0
3	189.2	159.35	2.202352068	6.607056203	9
6	207.1	150.4	2.177247836	13.06348702	36
9	224.4	141.75	2.151523068	19.36370761	81
12	240.2	133.85	2.126618376	25.51942051	144
15	256	125.95	2.100198172	31.50297258	225
18	265.7	121.1	2.083144143	37.49659458	324
21	282.6	112.65	2.051731196	43.08635512	441
$\Sigma x = 84$			$\Sigma y = 17.121471$	$\Sigma xy = 176.63959$	$\Sigma x^2 = 1260$

which gives

$$8a + 84b = 17.121471 \tag{5.58}$$

$$84a + 1260b = 176.63959 \tag{5.59}$$

which on solving for a and b gives

$$b = -0.0083 \text{ and } a = 2.2273$$

Since $b = -k/2.303$, $k = 0.0083 \times 2.303 = 0.0191 \text{ min}^{-1}$.

Example 5: Find the value of standard half electrode potential E^0 of Ag|AgCl using the following equation and data (Table 5.5):

$$E + \frac{2RT}{F} \ln m = E^0 + \frac{2.303RT}{F} \sqrt{m} \tag{5.60}$$

Table 5.5: LSC for standard electrode potential.

$\sqrt{m}$	0.05670	0.07496	0.09559	0.1158	0.1601	0.2322	0.3519
$E + \frac{2RT}{F} \ln m$	0.2256	0.2263	0.2273	0.2282	0.2300	0.2322	0.2346

Solution: The given equation may be interpreted as

$$y = a + bx$$

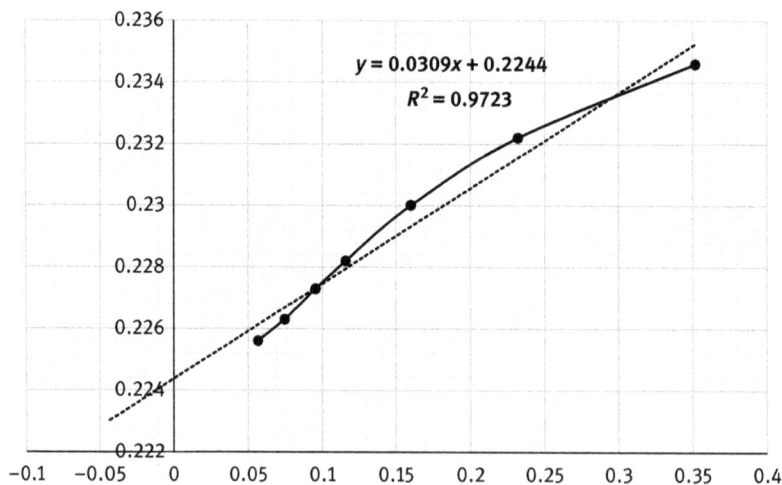

Figure 5.5: Variation of $E + \frac{2RT}{F} \ln m$ versus $\sqrt{m}$.

where intercept $a = E^0$, $b = 2.303RT/F$ and the data is plotted (Figure 5.5) which on extrapolation to y-axis gives $E^0_{Cl^-|AgCl|Ag} = 0.2244$ V.

5.5.2 Multivariate linear regression

Multiple linear regression (MLR) method is used when there are more than one variable is involved in influencing the dependent variable y. It may be represented as

$$y = a + bx_1 + cx_2 \qquad (5.61)$$

where x_1 and x_2 are the two independent variables while y is a dependent variable, b and c are the slopes or coefficients associated with x_1 and x_2 and a is a constant. The following example will clearly demonstrate the MLR.

Example 6: Using the method of MLR, find out the order and rate of reaction using the given kinetics data:

$$\text{Rate}(r) = k[CO]^m [Cl_2]^n \qquad (5.62)$$

Rate (mol L^{-1} s^{-1})	[CO] (mol L^{-1})	[Cl$_2$] (mol L^{-1})
0.012	0.1	0.1
0.00426	0.1	0.05
0.006	0.05	0.1
0.00213	0.05	0.05

Solution: Since this equation is not linearized and it needs to be linearized by taking log both sides as (Table 5.6)

$$\log r = \log k + m \log[CO] + n \log[Cl_2] \qquad (5.63)$$
$$\log y = \log k + m X_1 + n X_2 \qquad (5.64)$$
$$Y = K + m X_1 + n X_2 \qquad (5.65)$$

where $Y = \log r$, $K = \log k$, $X_1 = \log [CO]$ and $X_2 = \log [Cl_2]$

Table 5.6: Multiple linear regression for kinetics data.

N	Rate(y)	[CO] = x_1	[Cl$_2$] = x_2	X_1	X_2	$Y = \log y$
1	0.012	0.1	0.1	−1	−1	−1.92082
2	0.00426	0.1	0.05	−1.30103	−1	−2.37059
3	0.006	0.05	0.1	−1	−1.30103	−2.22185
4	0.00213	0.05	0.05	−1.30103	−1.30103	−2.67162

Here instead of using manual calculations, one may use spreadsheets to ease the calculation by using the LINEST function (explained in subsequent section). Upon solving,

$$K = \log k = 0.57329$$
$$\text{antilog}(0.57329) = k = 3.743 \left(\text{mol L}^{-1}\right)^{-3/2} \text{s}^{-1}$$

Similarly $\qquad m = 1 \text{ and } n = 1.494.$

Hence rate $= 3.743 \times (0.1)^1 \times (0.1)^{1.494}$
$\qquad$ rate $= 0.012 \text{ mol L}^{-1} \text{s}^{-1}$ which is same as given in the table; hence, the calculations are correct.

Example 7: The absorbance values of a mixture of unknown concentration of $KMnO_4$ and $K_2Cr_2O_7$ using colorimetry at two different wavelengths 372 and 290 nm were found to be 0.732 and 0.5888, respectively. Using the method of MLR, find the unknown concentrations of the two samples.
Solution: Using Lambert–Beer's law, we know $A = \varepsilon c l$.

Since both the components are non-interacting with each other, their absorbance are additive. Also if the path length of the cell is taken as unity, then

At 372 nm,

$$0.732 = \varepsilon_{KMnO_4}{}^{372} C_{KMnO_4} + \varepsilon_{K_2Cr_2O_7}{}^{372} C_{K_2Cr_2O_7} \qquad (5.66)$$

At 290 nm,

$$0.588 = \varepsilon_{KMnO_4}{}^{290} C_{KMnO_4} + \varepsilon_{K_2Cr_2O_7}{}^{290} C_{K_2Cr_2O_7} \qquad (5.67)$$

which after substituting the value of molar absorption coefficient becomes

$$0.732 = 805 C_{KMnO_4} + 1652 C_{K_2Cr_2O_7} \qquad (5.68)$$

$$0.588 = 1030 C_{KMnO_4} + 853 C_{K_2Cr_2O_7} \qquad (5.69)$$

These two equations are of the form $y = ax_1 + bx_2 + c$. In the above case, $c = 0$, which can be solved analytically (Table 5.7)

Table 5.7: Multiple linear regression for absorbance data.

Absorbance (Y)	X1	X2
0.588	805	1652
0.732	1030	853

and on solving using the LINEST function gives $C_{K2Cr2O7} = 2.76 \times 10^{-4}$ M, $C_{KMnO4} = 3.41 \times 10^{-4}$ M.

5.5.3 Polynomial regression

The curve fitting for second-order polynomial (e.g., equation of a parabola) is like solving three simultaneous linear equations. For instance, a second-order polynomial is given as

$$y = a + bx + cx^2 \tag{5.70}$$

where error is given as

$$e_i = y_{\text{measured}} - y_{\text{expected}} = y_i - (a + bx + cx^2) \tag{5.71}$$

$$\min S_r = \min e_i^2 = \min [y_i - (a + bx + cx^2)]^2 \tag{5.72}$$

To minimize the error, differentiate the square of the error with respect to a, b and c and equal it to zero:

$$\frac{dS_r}{da} = \frac{dS_r}{db} = \frac{dS_r}{dc} = 0 \tag{5.73}$$

$$\frac{dS_r}{da} = \sum_{i=1}^{n} [-2(y_i - (a + bx + cx^2))] = 0 \tag{5.74}$$

$$\frac{dS_r}{db} = \sum_{i=1}^{n} [-2x(y_i - (a + bx + cx^2))] = 0 \tag{5.75}$$

$$\frac{dS_r}{dc} = \sum_{i=1}^{n} [-2x^2(y_i - (a + bx + cx^2))] = 0 \tag{5.76}$$

which gives

$$na + \sum_{i=1}^{n} x_i b + \sum_{i=1}^{n} x_i^2 c = \sum_{i=1}^{n} y_i \tag{5.77}$$

$$\sum_{i=1}^{n} x_i a + \sum_{i=1}^{n} x_i^2 b + \sum_{i=1}^{n} x_i c = \sum_{i=1}^{n} x_i y_i \tag{5.78}$$

$$\sum_{i=1}^{n} x_i^2 a + \sum_{i=1}^{n} x_i^3 b + \sum_{i=1}^{n} x_i^4 c = \sum_{i=1}^{n} x_i^2 y \tag{5.79}$$

$$\begin{bmatrix} n & \sum x_i & \sum x_i^2 \\ \sum x_i & \sum x_i^2 & \sum x_i^3 \\ \sum x_i^2 & \sum x_i^3 & \sum x_i^4 \end{bmatrix} \begin{bmatrix} a \\ b \\ c \end{bmatrix} = \begin{bmatrix} \sum y_i \\ \sum x_i y_i \\ \sum x_i^2 y_i \end{bmatrix} \tag{5.80}$$

These normal equations can now be solved for the coefficients a, b and c using matrices. Similarly higher order polynomials can also be used to fit the given data points into the equation.

Example 8: The osmotic pressure values at various concentrations of a polymer in a solution at 298 K are as follows:

ρ (g cm^{-3})	0.0200	0.0150	0.0100	0.0075	0.0050	0.0025
Π (atm)	0.0117	0.0066	0.0030	0.00173	0.0009	0.00035

Using the LSC fitting method as per equation $y = a + bx + cx^2$ where $y = \Pi/\rho$, $x = \rho$ and $a = RT/M$, calculate M.
Solution: The given equation is for the the van't Hoff equation for a polymer for which the normal equations 5.77–5.79 may be written

$$na + \sum_{i=1}^{n} x_i b + \sum_{i=1}^{n} x_i^2 c = \sum_{i=1}^{n} y_i \tag{5.77}$$

$$\sum_{i=1}^{n} x_i a + \sum_{i=1}^{n} x_i^2 b + \sum_{i=1}^{n} x_i c = \sum_{i=1}^{n} x_i y_i \tag{5.78}$$

$$\sum_{i=1}^{n} x_i^2 a + \sum_{i=1}^{n} x_i^3 b + \sum_{i=1}^{n} x_i^4 c = \sum_{i=1}^{n} x_i^2 y \tag{5.79}$$

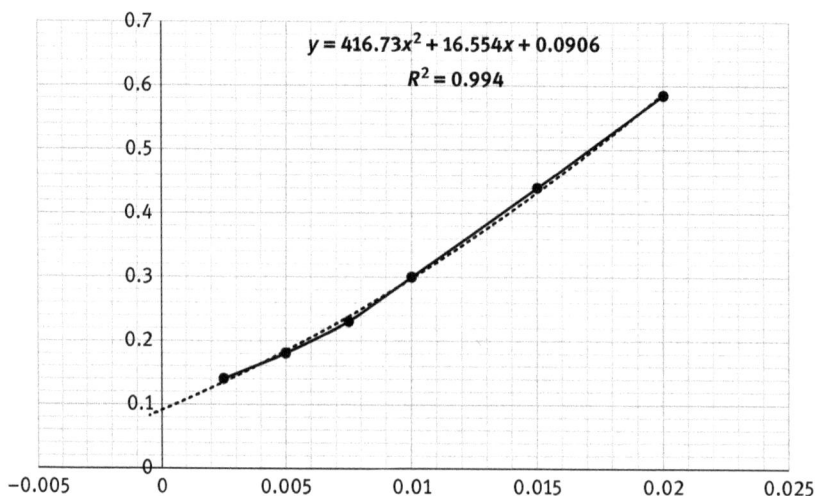

$$y = 416.73x^2 + 16.554x + 0.0906$$
$$R^2 = 0.994$$

Figure 5.6: Van't Hoff equation for a polymer.

The limiting value of Π/ρ at infinite dilution can be determined by plotting Π/ρ versus ρ and determining the intercept when $\rho \to 0$ (Figure 5.6). Therefore, $M = \frac{RT}{(\Pi/\rho)_{\rho \to 0}}$. Tabulating the required data in Table 5.8 as that gives

Table 5.8: Linear regression data.

$\rho = x$	Π	$\Pi/\rho = y$	x^2	x^3	x^4	xy	x^2y
0.02	0.0117	0.585	0.0004	0.000008	0.00000016	0.0117	0.000234
0.015	0.0066	0.44	0.000225	0.000003375	5.0625E-08	0.0066	0.000099
0.01	0.003	0.3	0.0001	0.000001	0.00000001	0.003	0.00003
0.0075	0.00173	0.230666667	0.00005625	4.21875E-07	3.16406E-09	0.00173	0.000012975
0.005	0.0009	0.18	0.000025	0.000000125	6.25E-10	0.0009	0.0000045
0.0025	0.00035	0.14	0.00000625	1.5625E-08	3.90625E-11	0.00035	0.000000875
$\Sigma x = 0.06$		$\Sigma y = 1.8756666$	$\Sigma x^2 = 0.00081$	$\Sigma x^3 = 1.293E-05$	$\Sigma x^4 = 2.244E-07$	$\Sigma xy = 0.0242$	$\Sigma x^2y = 0.0003813$

$$6a + 0.06b + 0.0008125c = 1.875666667 \qquad (5.81)$$

$$0.06a + 0.0008125b + 1.29375E - 05c = 0.02428 \qquad (5.82)$$

$$0.0008125a + 1.29375E - 05b + 2.2445E - 07c = 0.0003813 \qquad (5.83)$$

which on solving gives

$$a = 0.0906, \quad b = 16.554 \text{ and } c = 416.73$$

Since $a = RT/M$,
then $M = RT/a = 27,346.26 \text{ gmol}^{-1}$.

LSC fitting helps in trend analysis if one extrapolates the curve to predict what should be the behaviour or value of observable at any random point of interest. It compares the given measured data with the standard mathematical models (equations). It is a standard tool that is used in spreadsheets in terms of trendline that is derived from the LSC fitting itself.

5.6 Regression and correlation in EXCEL

Spreadsheets uses a set of statistical models and parameters called analysis of variance (ANOVA) to calculate statistical parameters including correlation and regression by using some keywords stored in the spreadsheets.

(1) To find the **correlation** one may enter the data in the two tables for which the relationship is to be determined, then in any blank cell, type CORREL (Array1, Array2) and press enter (Table 5.9).

Table 5.9: Correlation in spreadsheet.

	A	B
	Concentration	Absorbance
1		
2	0.01	0.29
3	0.02	0.35
4	0.03	0.48
5	0.04	0.6
6	0.05	0.71
	= CORREL(A2:A6,B2:B6)	

The result obtained is 0.9945 which is identical when using the Pearson correlation formula for which the keyword is PEARSON (Array1, Array2). The square of correlation coefficient is nothing but regression coefficient which can be verified either graphically or using LSC fitting since $A = \varepsilon cl$, linear relationship exists between absorbance and concentration having a slope of εl, with no intercept.

(2) To find the slope and intercept between the data points having linear relationship, **LINEST** function is used which uses the LSC method only.

For intercept, the syntax is (Table 5.10)

=INTERCEPT(known y's, known x's)
=INDEX(LINEST(known y's, known x's),2)
=INDEX(LINEST(B2:B6,A2:A6),2)

Table 5.10: LINEST function to find slope and intercept in spreadsheets.

	A	B
1	Concentration	Absorbance
2	0.01	0.29
3	0.02	0.35
4	0.03	0.48
5	0.04	0.6
6	0.05	0.71
	=INTERCEPT(B2:B6,A2:A6)	=SLOPE(B2:B6,A2:A6)

Likewise for slope, the syntax is

=SLOPE(B2:B6,A2:A6)
=INDEX(LINEST(B2:B6,A2:A6),1)

The 1 and 2 status in the syntax refers to the slope and intercept. The LINEST function returns the statistics for a line by using the method of LSC. It calculates the value of slope and intercept that best fits the given data. Not only LINEST function but SLOPE as well as the INTERCEPT function can also be used to evaluate slope and intercept, respectively. The results obtained are same as obtained by solving the equation used earlier.

LINEST function is most useful when there are two or more independent variable, also called MLR (Table 5.11). In such cases, the data is fed in the reverse order of the independent variable. For example, there are two independent variable x_1 and x_2 and one dependent variable (y), then after the data is fed, select three cells continuously in a row (A8,B8,C8) and enter the formula in the formula bar above and press **Ctrl + Shift + Enter**.

Its syntax is **LINEST(y's, x's(reverse order))**

Table 5.11: LINEST function for multiple linear regression.

$f(x)$	=LINEST(C2:C6, A2:B6)		
	A	B	C
1	$x1$	$x2$	Y
2	2	3	18
3	3	5	26
4	4	7	34
5	5	9	42
6	6	13	56
7	Slope2	Slope 1	Intercept
8	3	2	5

Hence by plugging the slope and intercept values, the equation may be written as

$$y = 3x_2 + 2x_1 + 5 \tag{5.84}$$

which can be verified by putting the known values of x_1 and

$$x_2 y = (2 \times 5) + (3 \times 9) + 5$$

$y = 42$ which is correct. Hence for any given value of x_1 and x_2, y can be calculated.

LINEST function can be used to calculate statistics not only for linear functions but for polynomial, exponential and logarithmic functions as well. **LOGEST** function is used to calculate the best exponential curve for the data while **GROWTH** function is used for exponential curve.

TREND just like LINEST also performs regression but does not return the value of slope and intercept in the spreadsheet but based on the results, it predicts the results (y values) for the new x values. Its syntax is

TREND(known_y's,known_x's) or
TREND(known_y's,known_x's, unknown x's value

FORECAST is similar to TREND function except that syntax is a bit different. While FORECAST can take only one predictor while TREND can do multiple predictors. These function can also be reached using the data analysis tab add ins.

(3)   To find the regression, one may use data analysis and select regression (Figure 5.7).

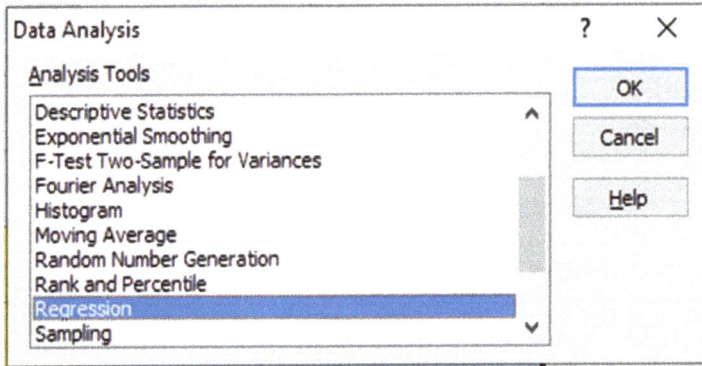

Figure 5.7: Data analysis tab.

A new window would appear asking to select the data range (Figure 5.8).

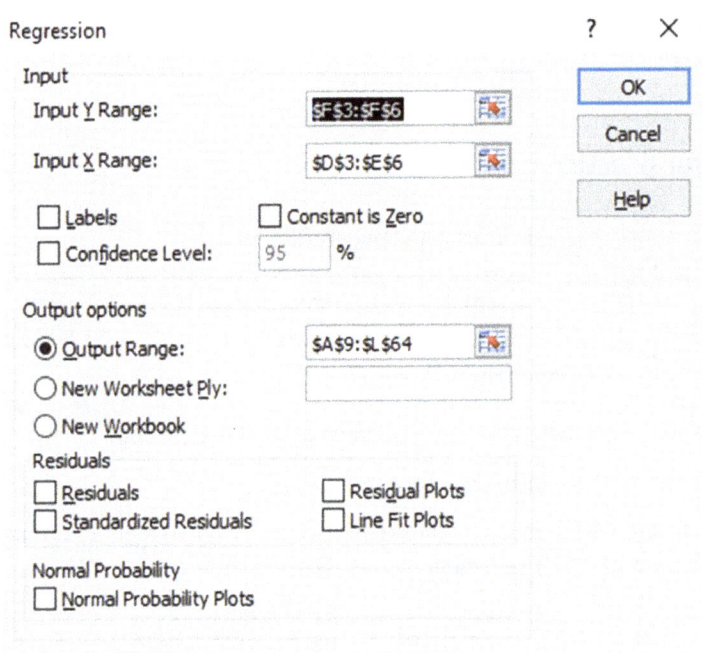

Figure 5.8: Selection of data in regression tab.

Once the data range is selected and the range of output is selected in a spreadsheet, the output would be displayed as shown in Figure 5.9.

SUMMARY OUTPUT

Regression statistics	
Multiple R	1
R square	1
Adjusted R square	1
Standard error	2.77556E−16
Observations	4

ANOVA

	df	SS	Ms	F	Significance F
Regression	2	0.292913593	0.146457	1.9E+30	5.1284E−16
Residual	1	7.70372E−32	7.7E−32		
Total	3	0.292913593			

	Coefficients	Standard error	t stat	P-value	Lower 95%	Upper 95%	Lower 95.0%	Upper 95.0%
Intercept	0.573290316	1.5066E−15	3.81E+14	1.67E−15	0.57329032	0.5732903	0.57329032	0.57329032
X Variable 1	1.49410907	9.2202E−16	1.62E+15	3.93E−16	1.49410907	1.4941091	1.49410907	1.49410907
X Variable 2	1	9.2202E−16	1.08E+15	5.87E−16	1	1	1	1

Figure 5.9: ANOVA in spreadsheet.

This analysis is based on **ANOVA**, which is a very powerful tool that may disburse a lot of information on statistical parameters and helps in constructing useful statistical model if the range of data is correct.

5.7 Exercises for practice

1. The viscosity of a hydrocarbon varies with temperature as

$t(°C)$	0	20	40
η (mP)	2.84	2.33	1.97

Determine the activation energy using LSC method as per the equation

$$\ln(\eta/\eta^0) = \ln(A/\eta^0) + \frac{E}{RT}$$

2. Using the LSC fitting, fit the following molar heat capacity for oxygen according to the equation $y = a + b(T/K) + c(T/K)^2$

T (K)	298	398	498	598	698
C_p (J K^{-1} mol^{-1})	29.38	30.24	31.07	31.88	32.67

3. Using the given data for decomposition of NO_2 (g), fit the data using LSC method for the equation $y = a + bx$ where $y = \log\{[NO_2]/\text{moldm}^{-3}\}$ and $x = t$ (s). If $b = -k/2.303$, find the rate constant k.

t (s)	0	200	400	600	800	1,000	1,200
$\dfrac{[NO_2]}{(\text{mol dm}^{-3})}$	0.250	0.223	0.198	0.174	0.152	0.134	0.120

4. The following data was obtained for the optical activity of a solution:

x	0	0.1	0.3	0.45	0.6	0.78	0.83	0.92
y	0	0.01	0.09	0.2025	0.36	0.6084	0.6889	0.8464

Using the LSC fitting method of the form $y = a + bx$, find the rate constant k, where $b = -k/2.303$, where $y = \log\{(\theta_\alpha - \theta_t)/(\theta_\alpha - \theta_0)\}$ and $x = t$ (min).

5. Using the LSC, fit the following data in a second-order polynomial

t (min)	0.0	7.2	36.8	46.8	68.0	a
θ (degree)	24.1	21.4	12.4	10.0	5.5	−10.7

6. Calculate K_d for the distribution of I_2 between CCl_4 and water (using LSC) for the given data.

C_{org} (mol dm^{-3})	1.8×10^{-3}	1.2×10^{-2}	1.75×10^{-2}	2×10^{-2}	2.5×10^{-2}
C_{aq} (mol dm^{-3})	1×10^{-4}	1.5×10^{-4}	2×10^{-4}	2.4×10^{-4}	3×10^{-4}

Chapter 6
Numerical integration

6.1 Introduction

Numerical integration refers to the approximate integration techniques used to evaluate integral where explicit function is unknown. There could be many ways in calculus to approximate the integral but numerical integration involves integration using numerical techniques. The idea behind this numerical technique is the approximation of a function which is continuous with discrete points. More simply, it involves finding the area under the curve where the curve represents the function (Figure 6.1). There are two techniques for numerical integration namely Newton–Cotes formula (where equal intervals are involved) and Gauss quadrature (where unequal intervals are present). Curve fitting is also a method of numerical integration which involves fitting a function to discrete points and then analytically integrating the function.

When the function is unknown, the discrete equally spaced data points are used to approximate the function by fitting them in a polynomial of nth order in Newton–Cotes formula. When the functions are known analytically or when there are unequal spaced sub-intervals, then Gaussian quadrature formula is used to calculate Gaussian quadrature at the selected abscissa to give the accurate approximations.

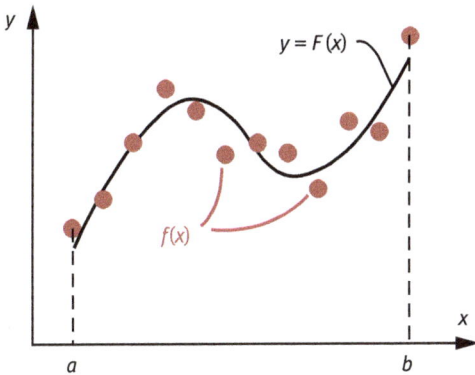

Figure 6.1: Definite integral of a function is the area under the curve.

Gauss quadrature involves three types of formulas namely Gauss–Legendre formula, Gauss–Chebyshev formula and Gauss–Hermite formula. In the context of numerical integration, this unit has been restricted to the most commonly used Newton–Cotes formula only, assuming the data is equally spaced.

https://doi.org/10.1515/9783111334448-006

6.2 Newton–Cotes formula

The Newton–Cotes closed formulas or Newton–Cotes quadrature formulas (named after Isaac Newton and Roger Cotes) refers to the set of formulas for numerical integration based on evaluation of integral having equally spaced intervals. They are the most common numerical methods of integration. Newton-closed formulas are referred to when the data points are known from the beginning to the end while Newton–Cotes open formulas are used when the integral limits are beyond data range. Since most definite integrals in sciences involve closed intervals, Newton-Cotes closed formulas are used. The Newton–Cotes closed formulas include
(1) Trapezoidal rule
(2) Simpson's 1/3rd rule
(3) Simpson's 3/8 rule
(4) Boole's rule

While trapezoidal rule approximates the integral using a straight line, Simpson's rule uses a curve or polynomial (Figure 6.2). Each method gets better by using better approximations of higher degree polynomial.

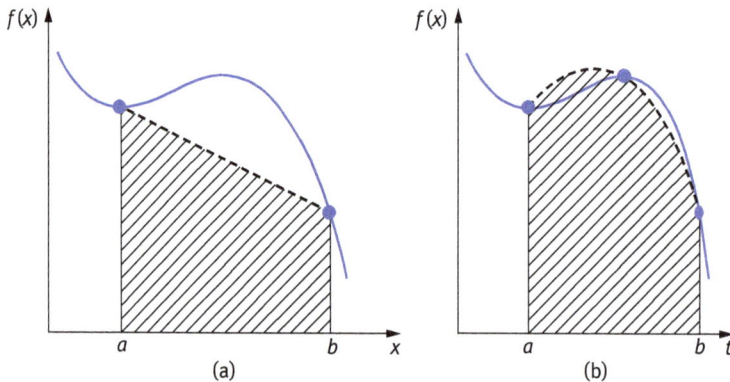

Figure 6.2: Approximating integral by a straight line (a) and a curve (polynomial) (b).

6.2.1 Riemann sum

This method is not a part of Newton–Cotes formula yet it is an important component when discussing integration. Riemann sum method is the method of dividing the area under the curve into various geometrical shapes like rectangle, trapezium, and so on and then adding the area of each component that gives the area under the curve. Rectangular sum is the method in which the area under the curve is divided into rectangle segments of which each area is calculated and integrated.

In order to calculate the area under the curve between the interval a and b (Figure 6.3), the interval $[a, b]$ is divided into n sub-intervals of width h as

$$h = \frac{b-a}{n} \tag{6.1}$$

by constructing a series of rectangles of width h and length $f(x)$. The sub-intervals are named as x_0, x_1, x_2, x_3, x_4 and x_5 (here $n = 5$).

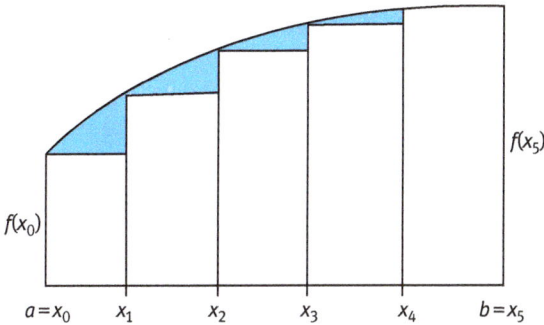

Figure 6.3: Riemann sum (underestimating the area under the curve).

The area under the curve can be approximated by adding the area of all the rectangles inside the curve as

$$\text{Area} = f(x_0)(x_1 - x_0) + f(x_1)(x_2 - x_1) + f(x_2)(x_3 - x_2) + f(x_3)(x_4 - x_3) + f(x_4)(x_5 - x_4) \tag{6.2}$$

$$\text{Area} = \{f(x_0) + f(x_1) + f(x_2) + f(x_3) + f(x_4)\}\Delta x \tag{6.3}$$

$$\text{Area} = \sum_{i=0}^{4} f(x_i)\Delta x \tag{6.4}$$

$$\text{Area} = \sum_{i=0}^{n-1} f(x_i)\Delta x \tag{6.5}$$

$$\text{Area} = \int_{0}^{n-1} f(x)dx \tag{6.6}$$

Here Δx is replaced by dx and Σ is replaced by $\int$ (integrand) to find the area under the curve. The earlier method will underestimate the area under the curve. There may be another way also to look at the area under the curve (Figure 6.4).

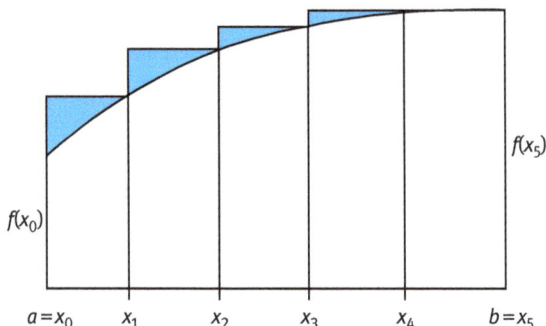

Figure 6.4: Riemann sum (overestimating the area under curve).

The formula and the method remains the same but this way the area under the curve is overestimated. Hence, Riemann's sum is a very vague method to find the integral. Clearly, this method seems to either underestimates or overestimates the integral. Hence, to account for better area, trapezoidal rule was introduced.

6.2.2 Trapezoidal rule

In mathematics or more specifically numerical analysis, the trapezoidal rule is a method used for approximating the definite integral $I = \int_a^b f(x)$. It is the simplest method for numerical integration where the interval points are connected by a chord to form a trapezium which is then often divided into multiple trapeziums whose total area is the approximation to the integral under consideration. This rule is mainly based on the Newton–Cotes formula of first order which states that one can find the exact value of the integral as an nth-order polynomial. The function $f(x)$ may not necessarily be a polynomial; it could be a linear function as well. The trapezoidal rule works by approximating the region under the graph of the function $f(x)$ as a trapezoid and calculates its area. This formula is also called Newton–Cotes two-point formula. Alike in Riemann sums, the area under the curve in trapezoidal rule is divided into several segments which are assumed to have trapezium shape. The area of each trapezoid is calculated and integrated. Since area of trapezium is given as

$$\text{Area of trapezium} = 1/2 \times (\text{sum of parallel sides}) \times \text{distance between them}$$

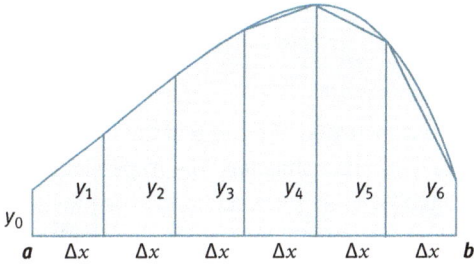

Figure 6.5: Trapezoidal rule.

If the area under the curve is divided into n number of trapezium (number of intervals) then according to the figure area (Figure 6.5)

$$\text{Area} = \frac{1}{2}(y_0 + y_1)\Delta x + \frac{1}{2}(y_1 + y_2)\Delta x + \frac{1}{2}(y_2 + y_3)\Delta x + \cdots + \frac{1}{2}(y_{n-1} + y_n)\Delta x \qquad (6.7)$$

Here distance between parallel sides is Δx.

$$\text{Area} = \left(\frac{y_0}{2} + y_1 + y_2 + y_3 + \cdots + y_{n-1} + \frac{y_n}{2}\right)\Delta x \qquad (6.8)$$

where

$$\Delta x = \frac{b-a}{n} = h$$

If we are given a function $f(x)$ which is continuous between the intervals a and b, then the interval is divided into n sub-intervals, so we can write

$$y_0 = f(a) \qquad (6.9)$$

$$y_1 = f(a + \Delta x) \qquad (6.10)$$

$$y_2 = f(a + 2\Delta x) \qquad (6.11)$$

$$y_{n-1} = f(a + i\Delta x) \qquad (6.12)$$

$$y_n = f(b) \qquad (6.13)$$

Here $i = n-1$ (n refers to the number of intervals)

$$\text{Area} = \int_a^b f(x) \approx \Delta x\left(\frac{y_0}{2} + y_1 + \cdots + y_{n-1} + \frac{y_n}{2}\right) \qquad (6.14)$$

$$\text{Area} = \int_a^b f(x) \approx \Delta x\left\{\left(\frac{y_0 + y_n}{2}\right) + (y_1 + y_2 + \cdots + y_{n-1})\right\} \qquad (6.15)$$

$$\text{Area} = I = \int_a^b f(x) \approx h\left\{ \left(\frac{f(a)+f(b)}{2}\right) + (f(a+ih)) \right\} \tag{6.16}$$

It is evident that as the number of intervals (n) is increased, the width of the interval decreases (Figure 6.6). This increases the accuracy of approximation. The trapezoidal rule results can be improvised by increasing the number of intervals and decreasing the step size (h) but it adds to the overall round off error. To evade this situation, Simpson's rule was devised by using higher order polynomial to approximate the function.

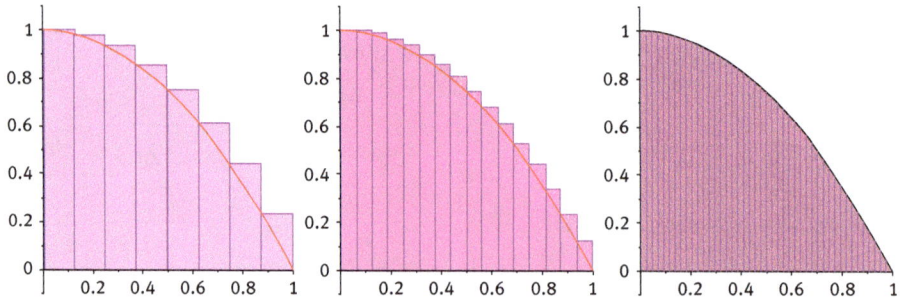

Figure 6.6: Increasing the sub-intervals (decreasing h).

Example 1: Using trapezoidal rule (Table 6.1), find the value of $\int f(x)dx = \int_0^1 (5x - 2x^2)dx$ using 10 intervals.

Solution: Here the limits of the integral a and b are given as 0 and 1, respectively, $n = 10$

$$h = \frac{b-a}{n} = \frac{1-0}{10} = 0.1$$

Table 6.1: Trapezoidal rule for evaluating integral.

	x_0	x_1	x_2	x_3	x_4	x_5	x_6	x_7	x_8	x_9	x_{10}
x	0	0.1	0.2	0.3	0.4	0.5	0.6	0.7	0.8	0.9	1.0
$f(x)$	0	0.48	0.92	1.32	1.68	2.0	2.28	2.52	2.72	2.88	3.0

So using the formula as above we have

$$I = 0.1\left\{ \frac{(0+3)}{2} + 0.48 + 0.92 + 1.32 + 1.68 + 2.0 + 2.28 + 2.52 + 2.72 + 2.88 \right\} \tag{6.17}$$

$$I = 1.83$$

The conventional way of integration gives $I = 1.8333$, so this method gives quite the accurate answer.

Example 2: Carry out the integration of the following equation

$$\frac{1}{N}\frac{dN}{du} = \int_a^b 4\pi \left(\frac{M}{2\pi RT}\right)^{3/2} \exp\left(\frac{-Mu^2}{2RT}\right) u^2 du \tag{6.18}$$

for O_2 gas with $du = (b-a)/N$ where $a = 0$, $b = 1{,}500$, $N = 15$ using all the three methods.

Solution: The given equation is Maxwell–Boltzmann's equation for distribution of speeds (Figure 6.7), for which the area under the curve represents the probability of finding the fraction of molecules having the said range of speeds which is always equal to 1. The same can also be verified using trapezoidal rule as well. For each molecular speed sub-interval, the ratio $\frac{1}{N}\frac{dN}{du}$ is tabulated (Table 6.2) and then integrating the area under the curve gives the probability of finding the molecule having that speed.

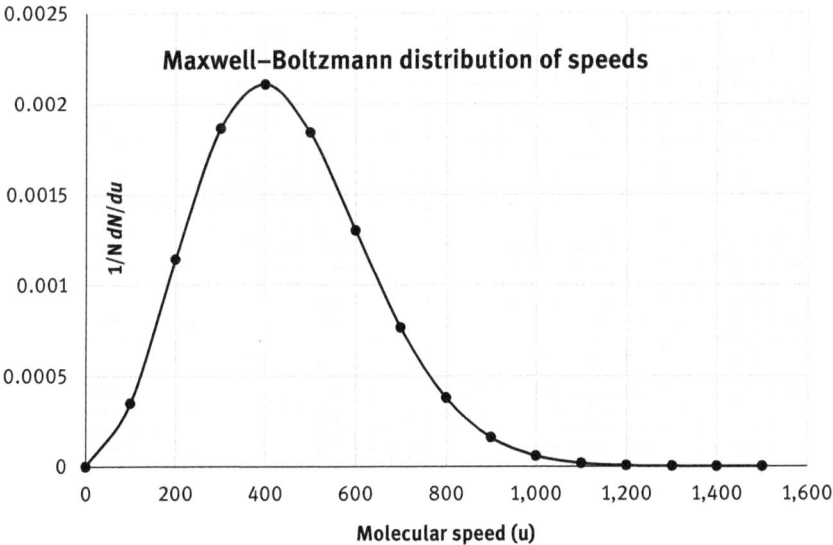

Figure 6.7: Maxwell distribution of molecular speeds for oxygen molecule.

Table 6.2: Trapezoidal rule for Maxwell–Boltzmann's distribution of speeds.

u (m/s)	dN/du (1/N)	u (m/s)	dN/du (1/N)
0	0	800	0.000380043
100	0.0003472	900	0.000160453
200	0.001144194	1,000	5.80738E–05
300	0.001864018	1,100	1.81046E–05
400	0.002108639	1,200	4.87863E–06

Table 6.2 (continued)

u (m/s)	dN/du (1/N)	u (m/s)	dN/du (1/N)
500	0.001842486	1,300	1.13936E–06
600	0.001303934	1,400	2.31087E–07
700	0.000766558	1,500	4.07714E–08

$$I = \frac{h}{2}[f(a) + f(b) + 2(f(a + (n-1)h)]$$ (6.19)

Since $$h = (1500 - 0)/15 = 100$$

$$I = 100[2.038 \times 10^{-8} + 9.999952 \times 10^{-3}]$$ (6.20)

$$I = 0.99999724 \approx 1$$

While the area turned out to be 0.9999972 using trapezoidal rule for speeds 0 to 1,500 m/s. One may also deduce the probability of finding the molecules in a particular speed range like from 400 to 1,000 m/s, it is found to be 0.553% or 55.3%.

Example 3: Calculate the change in entropy of nitrogen heated at constant pressure from 298 to 376 K where temperature variation of $C_{p,m}$ is given by the expression

$$C_{p,m}/JK^{-1}mol^{-1} = 27.296 + 5.23 \times 10^{-3}(T/K) - 0.042 \times 10^{-7}(T/K)^2$$ (6.21)

Solution: Since

$$\left(\frac{dS}{dT}\right)_p = \frac{C_{p,m}}{T}$$ (6.22)

Hence entropy change is given by (Table 6.3)

Table 6.3: Trapezoidal rule for finding the entropy.

T(K)	$C_{p,m}/T(JK^{-2}mol^{-1})$	T(K)	$C_{p,m}/T(JK^{-2}mol^{-1})$	T(K)	$C_{p,m}/T(JK^{-2}mol^{-1})$	T(K)	$C_{p,m}/T(JK^{-2}mol^{-1})$
298	0.0968	319	0.0908	340	0.0855	361	0.0808
301	0.0959	322	0.0900	343	0.0848	364	0.0802
304	0.0950	325	0.0892	346	0.0841	367	0.0796
307	0.0941	328	0.0884	349	0.0834	370	0.0790
310	0.0933	331	0.0877	352	0.0828	373	0.0784
313	0.0924	334	0.0870	355	0.0821	376	0.0778
316	0.0916	337	0.0862	358	0.0815		

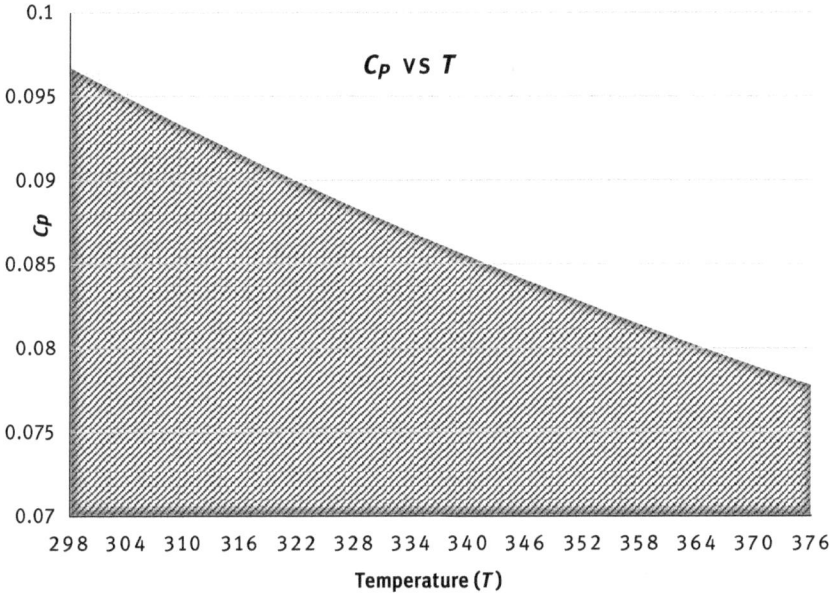

Figure 6.8: C_p versus T (variation of heat capacity with temperature).

$$\Delta S = \int \frac{C_p}{T} dT \qquad (6.23)$$

Therefore, plugging the given function of heat capacity C_p and integrating will give entropy change

$$\Delta S = \int \frac{C_{p,m}}{T} dT = \int \left(\frac{27.296}{T} + 5.23 \times 10^{-3}/K - 0.042 \times 10^{-7}(T/K^2) \right) dT \qquad (6.24)$$

The entropy change can be found by integrating the area under the curve of C_p/T versus T (Figure 6.8). *Using trapezoidal rule*, we get

$$\text{Area} = \frac{376 - 298}{26}[0.087325 + 2.164047]$$

$$= 6.754\, J\, K^{-1}\, mol^{-1}$$

While solving the integral by the conventional way of using limits, the change in entropy is found to be $6.75\, J\, K^{-1}$. Hence the results from trapezoidal rule is quite satisfactory because the number of intervals are large.

6.2.2.1 Error of trapezoidal rule

The error in the expression is derived using Taylor's expansion for $f(x)$ and is given by

$$E = -\frac{(b-a)^3}{12}f''(x) \qquad (6.25)$$

where
$$a < x < b.$$

where $f''(x)$ is the second derivative of $f(x)$ over the interval $[a,b]$. Hence, E is proportional to h^3.

6.2.3 Simpson's 1/3rd rule

Simpson's rule is the second type of Newton–Cotes method of second order for approximating the integral of function $f(x)$ using polynomial (e.g. parabolic arcs instead of the straight line used in trapezoidal rule). The Simpson's rule was named after Thomas Simpson to approximate the area under the curve and to account for the inaccuracy in Trapezoidal rule. While the trapezoidal rule works better for linear functions, Simpson's rule works very well for curves and polynomials. This method is more accurate than other numerical method of integration (Figure 6.9).

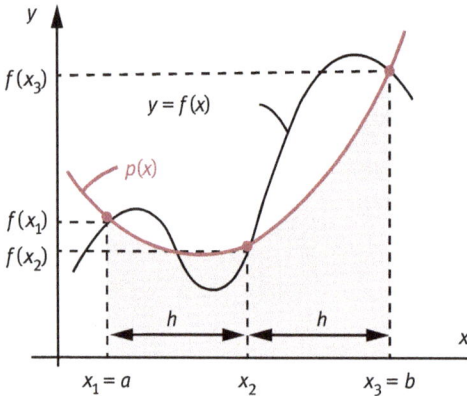

Figure 6.9: Approximating the data points with a parabolic function.

In Simpson's 1/3rd rule, a parabola is used to approximate each part of the curve. The area is divided into n equal segments of width Δx. Simpson's rule can be derived by approximating the Lagrange's polynomial. Here $P_i(x)$ is the polynomial of degree $\leq$ $n-1$ that passes through n points $(x_1, y_1), (x_2, y_2), \ldots, (x_n, y_n)$ and is given by

$$P(x) = \sum_{i=1}^{n} P_i(x) \tag{6.26}$$

$$P(x) = y_i \, \Pi_{k=1}^{n} \frac{x - x_k}{x_i - x_k} \tag{6.27}$$

$$l_i(x) = \Pi_{k=1}^{n} \frac{x - x_k}{x_i - x_k} \tag{6.28}$$

where

$$l_i(x) = \left(\frac{x - x_0}{x_i - x_0}\right)\left(\frac{x - x_1}{x_i - x_1}\right) \cdots \left(\frac{x - x_{i-1}}{x_i - x_{i-1}}\right)\left(\frac{x - x_{i+1}}{x_i - x_{i+1}}\right) \cdots \left(\frac{x - x_n}{x_i - x_n}\right) \tag{6.29}$$

Here the basic assumption made is that the area under the curve is divided into three equal parts as shown (Figure 6.10). This method involves approximating Lagrange's

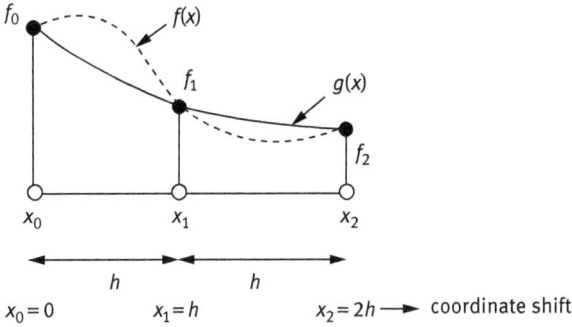

Figure 6.10: Approximating area using second-order Lagrange's polynomial P(x).

second-order polynomial where the interval is divided between x_0 and x_2 which gives the width of spacing between two intervals as

$$h = \frac{x_2 - x_0}{2} \qquad (6.30)$$

$$P(x) = f_0 L_0(x) + f_1 L_1(x) + f_2 L_2(x) \qquad (6.31)$$

where

$$L_0(x) = \left(\frac{x - x_1}{x_0 - x_1}\right)\left(\frac{x - x_2}{x_0 - x_2}\right) \qquad (6.32)$$

substituting the values of variables (x_0, x_1 and x_2)

$$L_0(x) = \frac{x^2 - 3hx + 2h^2}{2h^2} \qquad (6.33)$$

$$L_1(x) = \left(\frac{x - x_0}{x_1 - x_0}\right)\left(\frac{x - x_2}{x_1 - x_2}\right) \qquad (6.34)$$

$$L_1(x) = \frac{4hx + 2x^2}{2h^2} \qquad (6.35)$$

$$L_2(x) = \left(\frac{x - x_0}{x_2 - x_0}\right)\left(\frac{x - x_1}{x_2 - x_1}\right) \qquad (6.36)$$

$$L_2(x) = \frac{x^2 - hx}{2h^2} \qquad (6.37)$$

We know we can integrate function $f(x)$ as

$$I = \int_{x_0}^{x_2} f(x)dx \qquad (6.38)$$

Hence

$$I = \int_{x_0}^{x_2} \left\{ f_0 \left[\frac{x^2 - 3hx + 2h^2}{2h^2} \right] + f_1 \left[\frac{4hx + 2x^2}{2h^2} \right] + f_2 \left[\frac{x^2 - hx}{2h^2} \right] \right\} dx \qquad (6.39)$$

$$I = \frac{1}{2h^2} \left[f_0 \left(\frac{x^3}{3} - \frac{3hx^2}{2} + 2xh^2 \right) + f_1 \left(\frac{4hx^2}{2} - \frac{2x^3}{3} \right) + f_2 \left(\frac{x^3}{3} - \frac{hx^2}{2} \right) \right]_0^{2h} + E \qquad (6.40)$$

$$I = \frac{1}{2h^2} \left[f_0 \left(\frac{8h^3}{3} - \frac{12h^3}{2} + 4h^3 - 0 \right) + f_1 \left(8h^3 - \frac{16h^3}{2} \right) + f_2 \left(\frac{8h^3}{3} - \frac{4h^3}{2} \right) \right]_0^{2h} + E \qquad (6.41)$$

$$I = \frac{h}{3} [f_0 + 4f_1 + f_2] + E \qquad (6.42)$$

Since the final expression has a factor of 1/3, it is also called Simpson's 1/3 rule.

6.2.3.1 Composite integral Simpson's rule

If the interval of the integration is small and smooth (there is not much change in the function) then only two intervals like the earlier are sufficient to approximate the exact integral. But if they are not then in such cases, the intervals are further divided into many small sub-intervals to make the function smooth in such small intervals (Figure 6.11). The Simpson's rule is applied to each sub-interval and summed to give an integral which approximates for the entire interval. This is called Composite integral Simpson's rule is used which can be derived as

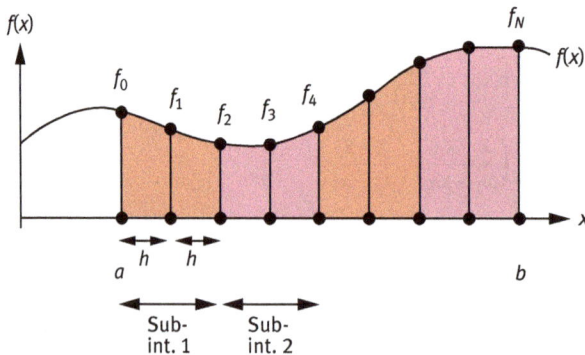

Figure 6.11: Subdividing the sub-intervals (composite Simpson's rule).

$$I = \frac{h}{3} [(f_0 + 4f_1 + f_2) + (f_2 + 4f_3 + f_4) + (f_4 + 4f_5 + f_6) + \cdots + (f_{N-4} + 4f_{N-3} + f_{N-2})$$
$$+ (f_{N-2} + 4f_{N-1} + f_N) + E \qquad (6.43)$$

$$I = \frac{h}{3}[(f_0 + 4f_1 + 2f_2 + 4f_3 + 2f_4 + 4f_5 + \cdots + 2f_{N-2} + + 4f_{N-1} + f_N) + E \qquad (6.44)$$

$$I = \frac{h}{3}\left[f(a) + f(b) + 4\sum_{i=1,2}^{N-1} f(a + ih) + 2\sum_{i=2,2}^{N-2} f(a + ih)\right] + E \qquad (6.45)$$

Here it is important to note that there are even number of intervals N while there are $N+1$ integration points which are odd. The Simpson's rule gives more accurate results than trapezoidal rule. Simpson's rule uses quadratic function to approximate the function over the given interval. The Simpson's rule demands a condition that N should be even and h must be a constant. Even if N is odd, h still must be constant.

The truncation error in the expression is given by

$$E = \frac{-1}{90}h^5 f''''(x) \qquad (6.46)$$

$$E = \frac{-(b-a)^5}{2880}f''''(x) \qquad (6.47)$$

where $f''''(x)$ is the fourth derivative of the function at a point in between $[a, b]$.

6.2.4 Simpson's 3/8 rule

Just like Simpson's 1/3 rule is based on quadratic approximation of the polynomial, when the polynomial is approximated by cubic interpolation, it is known as Simpson's 3/8 rule (Figure 6.12)

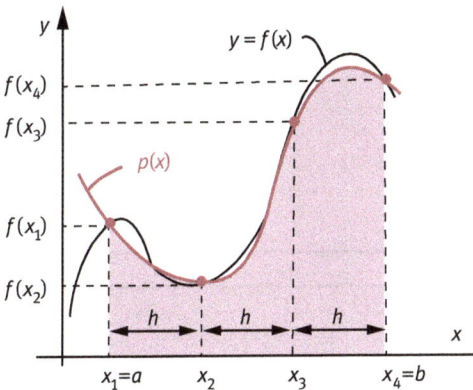

Figure 6.12: Simpson's 3/8 rule.

In this technique, cubic equation or higher order polynomial is used to approximate the data points. It uses third order Lagrange's polynomial which on integration gives

$$I = \int_{x_0}^{x_3} f(x)\,dx = \frac{3h}{8}[f(x_0) + 3f(x_1) + 3f(x_2) + f(x_3)] + E \qquad (6.48)$$

where the truncation error E is given by

$$E = \frac{-3h^5}{80} f^{(4)}(\xi) \qquad (6.49)$$

6.2.5 Boole's rule

When the integration is carried over by integrating the interval over five equally spaced sub-intervals, it is called Boole's rule which is given as

$$\int_{x_0}^{x_4} f(x)\,dx = \frac{2}{45} h(7f_0 + 32f_1 + 12f_2 + 32f_3 + 7f_4) + E \qquad (6.50)$$

where the truncation error E is given as

$$E = -\frac{8}{945} h^7 f^{(6)}(\xi) \qquad (6.51)$$

Example 4: Using the same data used in Example 2 for Maxwell distribution of speeds, calculate the area under the curve using Simpson's rule.

Solution: Use the Simpson's composite rule for more accurate result since there are large number of intervals and also there is marked difference between $f(x)$ values, hence

$$I = \frac{h}{3}\left[f(0) + f(1,500) + 4\sum f(\text{odd int}) + 2\sum f(\text{even int})\right] \qquad (6.52)$$

$h = (1500{-}0)/15 = 100$

$$I = \frac{100}{3}[0 + 4.0771 \times 10^{-8} + 4(0.004999958) + 2(0.004999994)] \qquad (6.53)$$

$I = 0.9999953 \approx 1$

Example 5: Using the relationship

$$\Delta S = S(T_2) - S(T_1) = \int_{T_1}^{T_2} \frac{C_p}{T}\,dT \qquad (6.54)$$

Calculate ΔS for heating of 1.00 mol of solid zinc from 20.0 to 300.0 K from the given data using both trapezoidal rule and Simpson's rule and compare the results.

Temperature(T) (in K)	$C_{p,m}$ ($JK^{-1}mol^{-1}$)	Temperature(T) (in K)	$C_{p,m}$ ($JK^{-1}mol^{-1}$)
20	18	180	162.8
40	44.56	200	181.5
60	64.43	220	200.9
80	80.88	240	221
100	96.15	260	241.8
120	111.6	280	263.1
140	127.8	300	284.7
160	144.9		

Solution: In this problem, we are given the temperature and the heat capacity at that temperature. According to the integral, we need to calculate C_p/T as y function which is tabulated (Table 6.4) and depicted graphically (Figure 6.13).

Table 6.4: Simpson's rule to find entropy change.

N	Temperature (T) (in K)	$C_{p,m}$ ($JK^{-1}mol^{-1}$)	$f(x) = C_{p,m}/T$
1	20	18	0.9
2	40	44.56	1.114
3	60	64.43	1.073
4	80	80.88	1.011
5	100	96.15	0.9615
6	120	111.6	0.93
7	140	127.8	0.9128
8	160	144.9	0.905625
9	180	162.8	0.9044
10	200	181.5	0.9075
11	220	200.9	0.9131
12	240	221	0.9208
13	260	241.8	0.93
14	280	263.1	0.9396
15	300	284.7	0.949

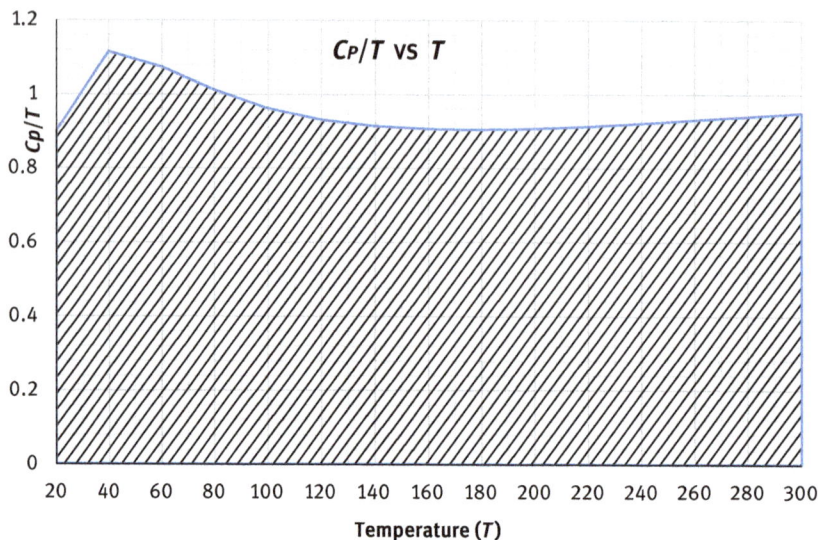

Figure 6.13: Variation of C_p/T with T (area under the curve gives entropy).

We can use both the Simpson and trapezoidal rules.

Using trapezoidal rule

$$\text{Area} = \frac{1}{2}\left[\frac{300-20}{14}(1.849+12.42335)\right] \tag{6.55}$$

$$I = 266.9565\,2\,\text{JK}^{-1}\,\text{mol}^{-1}$$

Using Simpson's rule

$$I = \frac{300-20}{14\times 3}[f(20)+f(300)+4(6.728525)+2(5.6948)] \tag{6.56}$$

$$I = 267.70025\,\text{J K}^{-1}\,\text{mol}^{-1}$$

When solving the function analytically, ΔS is found to be $253.9297\,\text{J K}^{-1}\,\text{mol}^{-1}$.

Example 6: What is the work done for a real gas if 1.00 mol of Cl_2 expands reversibly from 1 dm³ to 49 dm³ at 273 K where $a = 0.655$ MPa dm⁶ mol⁻² and $b = 0.055$ dm³ mol⁻¹ and $R = 8.314$ J K⁻¹ mol⁻¹.
Solution: We know that for real gases, van der Waal's equation of state for 1 mol of gas is given as

$$p = \frac{RT}{V-b} - \frac{a}{V^2} \tag{6.57}$$

and the work done is given by

$$w = -\int p\,dV = -\int\left(\frac{RT}{V-b} - \frac{a}{V^2}\right)dV \tag{6.58}$$

which upon integration gives

$$w = -RT \ln \frac{V_2 - b}{V_1 - b} - a \left(\frac{1}{V_2} - \frac{1}{V_1} \right) \tag{6.59}$$

For given volume, pressure is calculated using van der Waal's equation of state. The results are tabulated (Table 6.5) and pictorially depicted (Figure 6.14). Area under the pV curve is work done.

Table 6.5: Pressure–volume data for van der Waal's gas.

V (dm³)	P (kPa)	V (dm³)	p (kPa)	V (dm³)	p (kPa)
1	1.747	19	0.118	37	0.061
3	0.698	21	0.107	39	0.058
5	0.433	23	0.098	41	0.055
7	0.313	25	0.090	43	0.052
9	0.246	27	0.083	45	0.050
11	0.202	29	0.078	47	0.048
13	0.171	31	0.073	49	0.046
17	0.149	35	0.068		

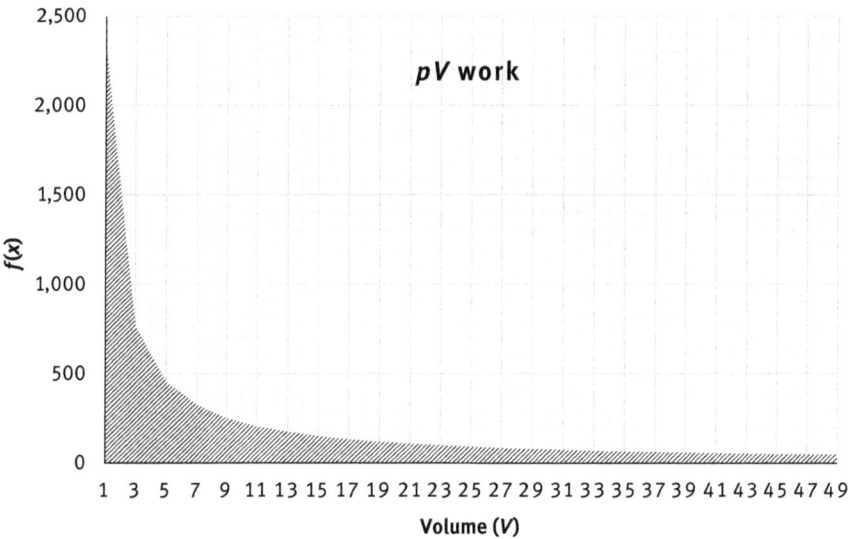

Figure 6.14: Area under the pV curve gives work done.

Using trapezoidal rule

$$w = \frac{1}{2} \times \left(\frac{49-1}{24}\right)[1.7929 + 2(3.4473)] \tag{6.60}$$

$$w = -8.687 \text{ kJ mol}^{-1}$$

Using Simpson's composite rule,

$$w = \frac{1}{3} \times \left(\frac{49-1}{24}\right)[1.7929 + 4(1.9567) + 2(1.4905)] \tag{6.61}$$

$$w = -8.400 \text{ kJ mol}^{-1}$$

The equation when solved analytically, work done is -8.319 kJ mol^{-1}.

In some examples earlier, there may be some considerable difference between the integral calculated analytically and numerically, since numerical methods are approximate methods and uses only given discrete data points as the basis of integration which may or may not be that accurate or deviate slightly from the original function in case of experimental data. Also the numerical methods use polynomials of nth order which may or may not exactly fit the data points. Yet we have observed that numerical integration is able to deliver satisfactory results where the original function is not available or is difficult to integrate.

Newton–Cotes open formulas
These methods do not use the end points or limits. There are three types of open formula:
(1)   Mid-point formula
(2)   Two-point formula
(3)   Three-point formula

Mid-Point formula
This rule doesn't use the end points rather use the mid point of the interval. The formula involves calculation of function at a single point i.e. at the mid point of the interval. Hence integral

$$\int_a^b f(x)\,dx \approx (b-a)f(x_m)$$

where x_m is the mid-point,

$$x_m = \frac{a+b}{2}$$

Two-point formula
The two-point formula uses two sub-intervals rather the original intervals. When the function $f(x)$ is continuous between the interval $[a,b]$ then using the two-point formula, hence integral is

$$\int_a^b f(x)\,dx \approx \frac{b-a}{2}(f(x_1) + f(x_2))$$

Three-point formula

In this method, the interval is divided into four sub-intervals such that

$$h = \frac{b-a}{4}$$

$$\int_a^b f(x)\,dx = \frac{4h}{3}(2f(x_1) - f(x_2) + 2f(x_3))$$

$$\int_a^b f(x)\,dx = \frac{b-a}{3}(2f(x_1) - f(x_2) + 2f(x_3))$$

where

$$x_1 = \frac{3a+b}{4}, x_2 = \frac{a+b}{2}, x_3 = \frac{3b+a}{4}$$

6.3 Exercises for practice

1. Integrate $\int_{10}^{20} x^3\,dx$ using trapezoidal rule.

2. Using Simpson's rule, integrate $\int_1^3 e^{2x^2}\,dx$.

3. What is the work done if 1 mol of a real gas expands reversibly isothermally from 1.001×10^{-3} m^3 to 0.100 m^3 using trapezoidal and Simpson's rule both where $a = 0.3640$ Pa m^6 mol^{-2} and $b = 4.267 \times 10^{-5}$ m^3 mol^{-1}.

4. Using Simpson's rule, Calculate the change in entropy of Nitrogen heated at constant pressure from 298 to 376 K using Simpson's rule:

$$C_{p,m}(\text{J K}^{-1}\,\text{mol}^{-1}) = 27.296 + 5.23 \times 10^{-3}(T/K) - 0.042 \times 10^{-7}(T/K)^2$$

5. Calculate the increase in enthalpy from 300 to 1,100 K using numerical integration given the following data

T (K)	300	500	700	900	1,100
C_p (J K^{-1} mol^{-1})	39.9	49.5	50.8	53.4	54.9

Chapter 7
Differential equations

7.1 Introduction

A differential equation is an equation having derivatives. There may be first-order, second-order and higher order differential equations. The order represents the highest order derivative in the equation. For example, $\frac{dy}{dx} = x\mathrm{Cos}x$ is the first-order differential equation, and $\frac{d^2y}{dx^2} = 2\mathrm{Sin}\,x + 5\mathrm{Cos}\,x$ is a second-order differential equation. Ordinary differential equations (ODEs) are the equations which have functions of one or more independent variable (usually x) and their derivatives. In fact vast class of first-order differential equations cannot be solved, so in such cases numerical methods to solve differential equations come to rescue. For that matter, we would restrict ourselves to linear, separable and exact differential equations with set of conditions. Even the differential equations that are separable cannot be solved explicitly. The numerical methods to solve differential equations can at least point out the direction in which the change is taking place, that is, by plotting a graph between y and x.

Numerical methods for differential equations are the numerical approximations to the solutions of differential equations but it does not imply the integral computation. Some differential equations which cannot be solved using conventional algebraic methods, separation of variables, integrating factors or any similar methods are solved using numerical methods. Even if the solution is obtained from the earlier mentioned algebraic means or methods, the solution may seem complicated. In such cases numerical methods give an approximate solution. The most conventional method of solving the ordinary differential equation numerically is Euler's method. The other method used is Runge–Kutta (RK) methods. Here what we are dealing with is ODEs i.e. Ordinary differential equations.

The differential equation are written as $dy/dx = f(x,y)$ where $y(0) = y_0$, that is, we are given an initial condition: at $x = 0$ and $y = 0$, which can also be written as x_0, y_0.

What are Cauchy–Euler differential equations?
Cauchy–Euler's differential equations also called Euler's equations are linear homogeneous differential equations with variable coefficients that can be solved explicitly. In general they can be represented as

$$a_n x^n y^{(n)}(x) + a_{n-1}x^{n-1}y^{(n-1)}(x) + \cdots + a_0 y(x) = 0 \qquad (7.1)$$

where a_n is constant, $y^{(n)}$ is the nth derivative of y w.r.t. x.

---

https://doi.org/10.1515/9783111334448-007

7.2 Euler's method

All differential equations cannot be solved implicitly. Euler's method is a numerical solution for first-order differential equations. The error in each step is directly proportional to the square of the step size.

This method uses the Taylor's series expansion to express the differential equation. The Taylor's expansion for a function is given by

$$f(x+h) = f(x) + \frac{h}{1!}\frac{dy}{dx} + \frac{h^2}{2!}\frac{d^2y}{dx^2} + \frac{h^3}{3!}\frac{d^3y}{dx^3} + \cdots \tag{7.2}$$

or

$$y(x+h) = y(x) + \frac{h}{1!}y'(x) + \frac{h^2}{2!}y''(x) + \frac{h^3}{3!}y'''(x) + \cdots \tag{7.3}$$

where h is the difference between two successive points like x and $x + h$. If the interval between different x values is small, i.e. h is small, then Euler's method approximates the expansion term in Taylor's series so the earlier expansion can be approximated as (ignoring higher order terms because they are very small)

$$y(x+h) \approx y(x) + hf(x,y) \tag{7.4}$$

where $f(x,y)$ is the derivative of y at (x_0, y_0), x_0 is the initial value of x and y_0 is the value of y at $x = x_0$ (i.e. the initial condition). In general, one may write

$$y_{n+1} = y_n + hf(x_n, y_n) \tag{7.5}$$

y_{n+1} is the next estimated value, y_n is the current value, h is the interval size and $f(x_n, y_n)$ is the derivative at $[x_n, y_n]$. Since Euler's method retains only first-order derivative terms from the Taylor's series expansion of the function, it is also called RK's method of first order.

A first-order differential equation may be considered as a derivative or slope of the function:

$$\frac{dy}{dx} = f(x,y) \tag{7.6}$$

The Euler's method uses the basis of initial value problem which suggests that as we calculate the next value of y_1 from initial y_0 value, this calculated value becomes the initial value for the next higher value of y_2.

Example 1: Using the Euler's method, solve the following differential equation:

$$\frac{dy}{dx} = 5 - 3\frac{y}{x} \tag{7.7}$$

Given $y(3) = 1$ and $h = 0.1$.

Solution: The initial conditions are given as $y(3) = 1$, that is, at $x = 3$, $y = 1$. Using eq. (7.7), dy/dx is calculated as

$$f(x, y) = \frac{dy}{dx} = 5 - 3\left(\frac{1}{3}\right) = 4 \tag{7.8}$$

Since

$$x_{n+1} = x_n + h = 3.1$$

likewise

$$y_{n+1} = y_n + hf(x, y), \text{ which gives } y_{n+1} = 1 + (0.1)(4) = 1.4$$

The remaining results are tabulated in Table 7.1 and depicted graphically in Figure 7.1. One may calculate the data points using the earlier differential equation.

Table 7.1: Euler's method to solve first-order ODE.

x	y	dy/dx
3	1	4
3.1	1.4	3.64516129
3.2	1.764516129	3.345766129
3.3	2.099092742	3.091733871
3.4	2.408266129	2.875059298
3.5	2.695772059	2.689338235
3.6	2.964705882	2.529411765
3.7	3.217647059	2.391096979
3.8	3.456756757	2.270981508

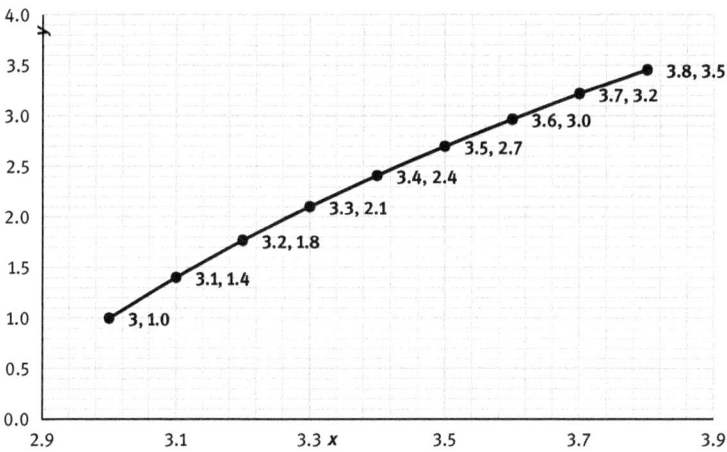

Figure 7.1: Solved ODE using Euler's method.

Thus this graph is the solution of the above differential equation.

Example 2: Using the solution of differential equation for first-order kinetics $y' = -ky$ with the initial condition of $[A_0] = 0.25$ mol dm^{-3}, determine the concentration $[A]$ at $t = 200, 400, 600, 800, 100$ and $1,200$ s where $k = 0.000622$ s^{-1}.

Solution: Here a first-order differential equation, Euler's method would be used as

$$\frac{d[A]}{dt} = -k[A] \tag{7.9}$$

where initial condition is $[A](t = 0) = [A_0] = 0.25$, $h = 200$ s

$$y_1 = A_{200} = A_0 + \Delta t.d[A]/dt \tag{7.10}$$

$$A_{200} = 0.25 + \left(200 \times \frac{dy}{dx}\right) \tag{7.11}$$

The results are further tabulated in Table 7.2 and depicted graphically in Figure 7.2.

Table 7.2: Euler's method to solve for kinetics data.

x	y	dy/dx
0	0.25	−0.0001555
200	0.218	−0.000136156
400	0.191	−0.000119218
600	0.167	−0.000104387
800	0.146	−9.14015E−05
1,000	0.128	−8.00312E−05
1,200	0.112	−7.00255E−05

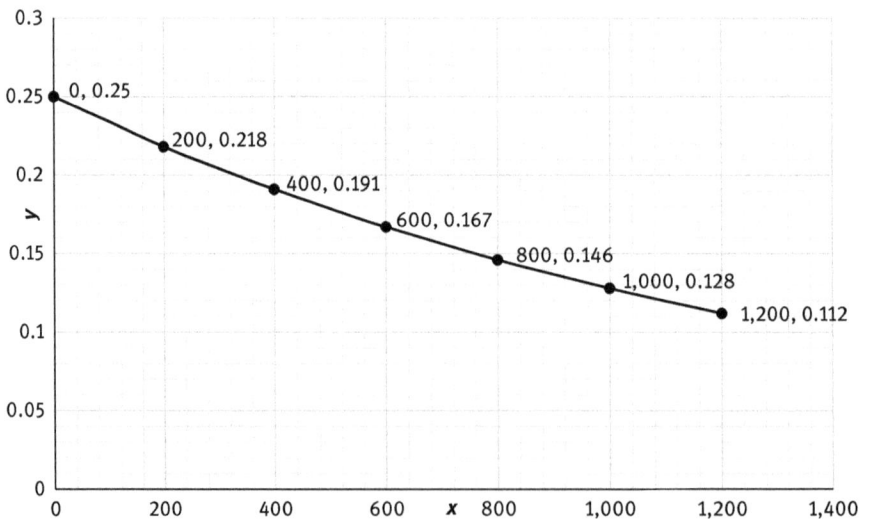

Figure 7.2: Euler's method for solving first-order kinetics data.

Its solution is given by

$$\ln y = -kx + C \tag{7.12}$$

where $C = \ln 0.25$ which gives the same results as found out by Euler's method while the general solution is

$$[A] = [A]_0 e^{-kt} \tag{7.13}$$

or

$$\ln[A] = \ln[A]_0 - kt \tag{7.14}$$

$$\ln[A] = C - kt \tag{7.15}$$

7.3 Runge–Kutta method

As compared to Euler's method which requires very small intervals to give accurate results, RK method produces better results in lesser steps. It is a numerical method to integrate a differential equation. RK methods are robust method which gives good numerical solution of differential equations if the step size is taken intelligently. RKs fourth-order algorithm is more popular due to their accuracy in the results.

The aim of RK method is same as of Euler's method, that is, to find the value of y at any other value of x if the initial values are given. Euler's method will be exact when the solution is a line while RK methods will be accurate if the solution is a polynomial of higher order. (Euler's method is same as RK1 method.) RK methods for higher order polynomial are more accurate and involve multi-stage calculations involving calculations at various slope values (e.g. RK3 and RK4 methods).

RK method provides good approximation to a function without having to differentiate the equation unlike Taylor's series (only if the point of interest is close enough). Taylor's method differentiates the function for each new term one intends to calculate. RK method aims to simulate as many steps of Taylor's series while evaluating the original function. RK method order 4 formula, also called RK4, is widely used in the fields of mechanics, climate models, aerodynamics, environmental studies models, drug delivery, ecosystem study and many other sub-branches of physics, chemistry and biology.

7.3.1 RK 2 or predictor–corrector method

RK method of order 2 is also called Heun's method or improved Euler's method. This numerical method is used to solve only first-order differential equations of the form

$$\frac{dy}{dx} = f(x, y) \tag{7.6}$$

It involves calculating the intermediate value and then use this value to calculate the final value. We know by Euler's method that

$$y_{n+1} = y_n + hf(x_n, y_n) \tag{7.16}$$

where the symbols have their usual meaning as described in the earlier methods. The Taylor's series is given as

$$y(x + h) = y(x) + \frac{h}{1!}y'(x) + \frac{h^2}{2!}y''(x) + \frac{h^3}{3!}y'''(x) + \cdots \tag{7.17}$$

If the first two terms of the Taylor's series are retained then it is nothing but Euler's method or RK's first-order method. But if the first three terms are retained, then the approximation would look like

$$y_{n+1} = y_n + hf(x_n, y_n) + \frac{h^2}{2!}f'(x_n, y_n) \tag{7.18}$$

And is called RK second-order method where $f'(x_n, y_n)$ is the double derivative of function at point x_n, y_n. This implies that second-order derivative is to be calculated which is very difficult to calculate using the differential equation. Instead RK wrote second-order method as

$$y_{n+1} = y_n + \left(\frac{k_1}{2} + \frac{k_2}{2}\right)h \tag{7.19}$$

where $k_1 = f(x_n, y_n)$ and $k_2 = f(x_n + h, y_n + k_1 h)$.

Here y_{n+1} is the final value while $y_n + k_1 h$ is the intermediate value.

The true error in the approximation is given by

$$E = \frac{h^2}{2!}f(x_n, y_n) + \frac{h^3}{3!}f'(x_n, y_n) + \cdots \tag{7.20}$$

which is nothing but the remaining terms of the Taylor's expansion series.

Example 3: Using the solution of second-order RK method of differential equation for kinetics, as $dy/dx = -ky^2$ with the initial condition of $y(t = 0) = 632$ mol dm^{-3} at $t = 3,6, 9,12, 15$ and 18 min where $k = 1.78 \times 10^{-5}$ dm^3 min^{-1}.

Solution: Here the initial condition is given as $x = 0$ and $y = 632$

$$\frac{dy}{dx} = -ky^2 \tag{7.21}$$

Using the second-order RK method of differential equation we have

$$y_{n+1} = y_n + \left(\frac{k_1}{2} + \frac{k_2}{2} \right) h \qquad (7.19)$$

where $k_1 = f(x_n, y_n)$ and $k_2 = f(x_n + h, y_n + k_1 h)$

$$k_1 = f(x, y) = \frac{dy}{dx} = -1.78 \times 10^{-5}(632)^2 = -7.109 \qquad (7.22)$$

Similarly $x_n + h = 0 + 3 = 3$ and $y_n + k_1 h = 632 + (-7.10)3 = 610.67$

$$k_2 = f(3, 610.67) = -1.78 \times 10^{-5}(610.67)^2 = -6.637 \qquad (7.23)$$

Hence, the results are further tabulated in Table 7.3.

Table 7.3: RK2 method for solving kinetics data.

x	y	k_1	k_2	x + h	$y + k_1 h$
0	632	-7.1097472	-6.6379541	3	610.670
3	611.38	-6.6533481	-6.2260079	6	591.418
6	592.06	-6.2395114	-5.8512118	9	573.340
9	573.92	-5.8631061	-5.5092340	12	556.334
12	556.86	-5.5197519	-5.1963557	15	540.305
15	540.79	-5.2056907	-4.9093698	18	525.173
18	525.62	-4.9176834	-4.6454985	21	510.865
21	511.27	-4.6529267	-4.4023266	24	497.314

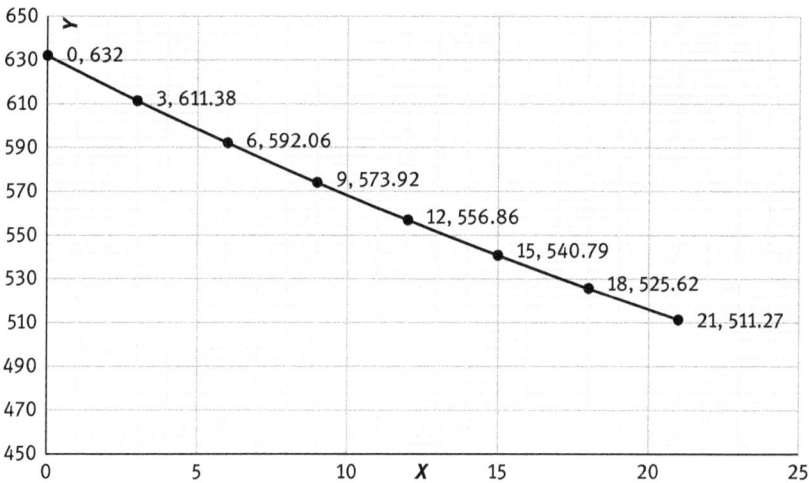

Figure 7.3: RK2 method for solving ODE.

The solution of the earlier differential equation is (Figure 7.3)

$$y = \frac{1}{kt + C} \tag{7.24}$$

where $C = 1/632$ using initial condition $y(t = 0) = 632$, hence

$$y = \frac{1}{kt + (1/632)} \tag{7.25}$$

which gives the same results.

7.3.2 RK3

RK3 is also known as RK method of order 3. In RK3 method, one more Taylor series term is included as compared to RK2 method. RK3 is comparatively more accurate than RK2 but less precise than RK4

$$y_{n+1} = y_n + \frac{1}{6}(k_1 + 2k_2 + k_3)h \tag{7.26}$$

where $k_1 = f(x_n, y_n)$, $k_2 = f(x_n + h/2, \ y_n + k_1 h/2)$ and $k_3 = f(x_n + h, \ y_n - k_1 h + 2k_2 h)$

7.3.3 RK4

Also known as RK method of order 4 is the most popular method to solve first-order differential equations. Apart from being the most accurate, it requires less computation time. In general, the RK formula for any $y(x + h)$ point can be written as

$$y_{n+1} = y_n + \frac{1}{6}(k_1 + 2k_2 + 2k_3 + k_4) \tag{7.27}$$

where $k_1 = hf(x, y)$, $k_2 = hf\left(x + \frac{h}{2}, y + \frac{k_1}{2}\right)$, $k_3 = hf\left(x + \frac{h}{2}, y + \frac{k_2}{2}\right)$, and $k_4 = hf(x + h, y + k_3)$

Originally the RK method was derived analytically rather from geometric mean point of view since from the formula, it is clear that it takes into account the geometric weight of k_i.

Example 4: Using RK method of order 4, solve the following differential equation using step size $h = 0.1$:

$$(1 + x^2)\frac{dy}{dx} + xy = 0 \tag{7.28}$$

$y(0) = 2, h = 0.1$

Solution: On rearranging, we have

$$\frac{dy}{dx} = \frac{-xy}{1 + x^2} \tag{7.29}$$

Using initial value problem, we have $y(0) = 2$, that is, at $x = 0, y = 2$,

$k_1 = hf(x, y) = 0.1f(0, 2) = 0,$

For, k_2, $x + \frac{h}{2} = 0 + \frac{0.1}{2} = 0.05$, $y + \frac{k_1}{2} = 2 + \frac{0}{2} = 2$

So $k_2 = hf\left(x + \frac{h}{2}, y + \frac{k_1}{2}\right) = 0.1f(0.05, 2) = -0.010$

For k_3, $x + \frac{h}{2} = 0 + \frac{0.1}{2} = 0.05$, $y + \frac{k_2}{2} = 2 + \frac{-0.010}{2} = 1.995$

$\quad k_3 = hf\left(x + \frac{h}{2}, y + \frac{k_2}{2}\right) = 0.1f(0.05, 1.995) = -0.010$

For k_4, $x + h = 0 + 0.1 = 0.1$, $y + k_3 = 2 + (-0.010) = 1.990$

$\quad k_2 = hf(x + h, y + k_3) = 0.1f(0.1, 1.990) = -0.020$

The rest of the results are tabulated in Table 7.4 and graphical solution is depicted in Figure 7.4.

Table 7.4: RK4 method to solve first-order differential equation.

x	y	dy/dx	k_1	$k_2(x)$	$k_2(y)$	k_2	$k_3(y)$	k_3	$k_4(x)$	$k_4(y)$	k_4
0.0	2.000	0.000	0.000	0.050	2.000	−0.010	1.995	−0.010	0.100	1.990	−0.020
0.1	1.990	−0.040	−0.004	0.150	1.988	−0.029	1.975	−0.029	0.200	1.961	−0.038
0.2	1.964	−0.081	−0.008	0.250	1.960	−0.046	1.941	−0.046	0.300	1.918	−0.053
0.3	1.923	−0.123	−0.012	0.350	1.917	−0.060	1.893	−0.059	0.400	1.864	−0.064
0.4	1.871	−0.166	−0.017	0.450	1.862	−0.070	1.836	−0.069	0.500	1.802	−0.072
0.5	1.810	−0.212	−0.021	0.550	1.799	−0.076	1.772	−0.075	0.600	1.735	−0.077
0.6	1.743	−0.259	−0.026	0.650	1.730	−0.079	1.704	−0.078	0.700	1.665	−0.078
0.7	1.674	−0.308	−0.031	0.750	1.658	−0.080	1.634	−0.078	0.800	1.595	−0.078
0.8	1.603	−0.359	−0.036	0.850	1.585	−0.078	1.564	−0.077	0.900	1.526	−0.076
0.9	1.532	−0.412	−0.041	0.950	1.512	−0.075	1.495	−0.075	1.000	1.458	−0.073
1.0	1.463	−0.466	−0.047	1.050	1.440	−0.072	1.427	−0.071	1.100	1.392	−0.069

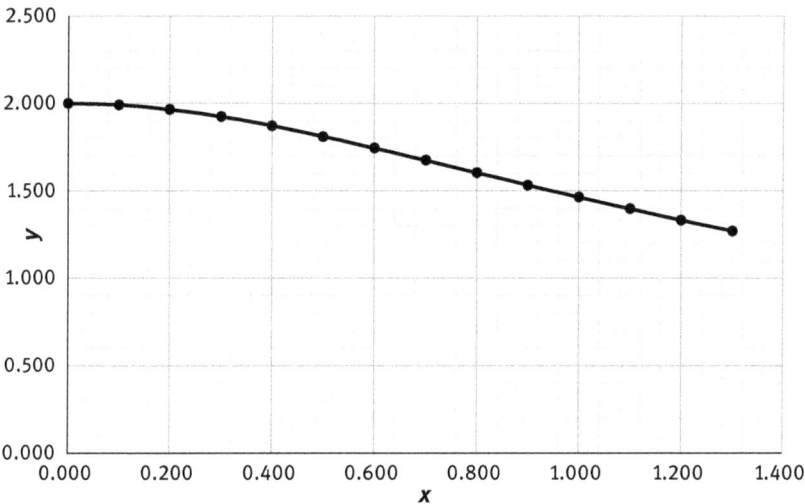

Figure 7.4: RK4 method to solve ODE.

Second-order differential equations can also be solved using the earlier numerical methods for solving differential equation. One such example of such equation is vibrating springs where an object of mass is held at the end of a spring (Hooke's law). According to the latter, the restoring force is given as $F = -kx$, where k is the spring constant. Also using Newton's law, $F = ma$ where m is the mass of the object held to the spring where acceleration can be written in terms of derivative of displacement of the spring from the equilibrium position. This gives a second-order differential equation as

$$m\frac{d^2x}{dt^2} = -kx \tag{7.30}$$

and the general solution is given by

$$x(t) = c_1\cos\omega t + c_2\sin\omega t \tag{7.31}$$

where

$$\omega = \sqrt{\frac{k}{m}}$$

For such second-order differential equations

$$m\frac{d^2y}{dx^2} + n\frac{dy}{dx} + ky = 0 \tag{7.32}$$

The equations can be written as two simultaneous coupled first-order equations by defining a new variable z as

$$\frac{dy}{dx} = z = f(x, y, z) \tag{7.33}$$

$$\frac{d^2y}{dx^2} = \frac{dz}{dx} = g(x, y, z) \tag{7.34}$$

Hence the earlier equation can be rewritten as

$$m\frac{dz}{dx} + nz + ky = 0 \tag{7.35}$$

This equation can also be solved using either Euler's method or RK4 method. For RK4 method, different coefficients are given as

$$k_0 = hf(x_0, y_0, z_0), \; l_0 = hg(x_0, y_0, z_0)$$
$$k_1 = hf\left(x_0 + \frac{h}{2}, y_0 + \frac{k_0}{2}, z_0 + \frac{l_0}{2}\right), \; l_1 = hg\left(x_0 + \frac{h}{2}, y_0 + \frac{k_0}{2}, z_0 + \frac{l_0}{2}\right)$$
$$k_2 = hf\left(x_0 + \frac{h}{2}, y_0 + \frac{k_1}{2}, z_0 + \frac{l_1}{2}\right), \; l_2 = hg\left(x_0 + \frac{h}{2}, y_0 + \frac{k_1}{2}, z_0 + \frac{l_1}{2}\right)$$
$$k_3 = hf(x_0 + h, y_0 + k_2, z_0 + l_2), \; l_3 = hg(x_0 + h, y_0 + k_2, z_0 + l_2)$$
$$y_1 = y_0 + \tfrac{1}{6}(k_0 + 2k_1 + 2k_2 + k_3), \; z_1 = z_0 + \frac{1}{6}(l_0 + 2l_1 + 2l_2 + l_3)$$

Example 5: A spring of 2 kg is stretched to length 0.7 m with a force of 25.6 N, which had the equilibrium length of 0.5 m then released with the initial velocity of zero. Find the position of mass at 0.7 s (Figure 7.5).

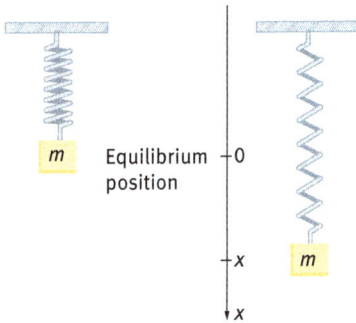

Figure 7.5: Hooke's law for a spring.

Solution: As we have learnt, the solution of the above type of second-order differential equation is given by

$$m\frac{d^2x}{dt^2} = -kx \tag{7.30}$$

According to Hooke's law

$$k(0.2) = 25.6$$
$$k = 128$$

Substituting the values of mass and spring constant as

$$2\frac{d^2x}{dt^2} + 128x = 0 \tag{7.36}$$

when solved analytically, the general solution is

$$x(t) = c_1\cos8t + c_2\sin8t \tag{7.31}$$

where

$$\omega = \sqrt{\frac{128}{2}} = 8$$

The initial condition of $x(t=0) = 0.2$, also the initial velocity is given as zero, that is, $x'(t=0) = 0$ or $dx/dt = 0$. Substituting the initial condition in 7.31

$$x(0) = c_1\cos0 + c_2\sin0 = 0.2 \tag{7.37}$$
$$x'(t) = 8c_1\sin8t + 8c_2\cos8t \tag{7.38}$$

Hence $c_1 = 0.2$. Also

At $t = 0$
$$x'(t=0) = 8c_1\sin0 + 8c_2\cos0 = 0 \tag{7.39}$$

Table 7.5: RK4 method to solve second-order differential equation.

x	y	z	k_0	k_1	k_2	k_3	l_0	l_1	l_2	l_3
0	0.2000	0.0000	0.0000	-0.0640	-0.0640	-0.1075	-1.2800	-1.2800	-1.0752	-0.8704
0.1	0.1394	-1.1435	-0.1143	-0.1590	-0.1407	-0.1527	-0.8922	-0.5263	-0.3836	0.0080
0.2	-0.0050	-1.5941	-0.1594	-0.1578	-0.1323	-0.1057	0.0318	0.5419	0.5368	0.8786
0.3	-0.1459	-1.0828	-0.1083	-0.0616	-0.0443	0.0048	0.9336	1.2801	1.1307	1.2170
0.4	-0.1984	0.0792	0.0079	0.0714	0.0701	0.1121	1.2699	1.2445	1.0413	0.8209
0.5	-0.1312	1.1896	0.1190	0.1610	0.1419	0.1514	0.8399	0.4592	0.3248	-0.0684
0.6	0.0148	1.5795	0.1580	0.1532	0.1279	0.0995	-0.0947	-0.6001	-0.5850	-0.9136
0.7	0.1514	1.0165	0.1016	0.0532	0.0369	-0.0123	-0.9691	-1.2944	-1.1393	-1.2054
0.8	0.1964	-0.1572	-0.0157	-0.0786	-0.0760	-0.1162	-1.2567	-1.2064	-1.0053	-0.7700

which gives $c_2 = 0$. Hence the solution we have is

$$x(t) = 0.2\cos 8t \qquad (7.40)$$

To find displacement at $t = 0.7$ s, $x(0.7) = 0.1551$. Likewise, the equation can be solved using the RK's method for solving differential equations (Figure 7.6) as shown in Table 7.5 using the above methodology.

Using the RK method, we got almost the same answer as 0.1514. Since they are approximate methods, one may not always get the same exact answer as that of analytical solution but quite close solution depending on the stability of the algorithm for the given problem.

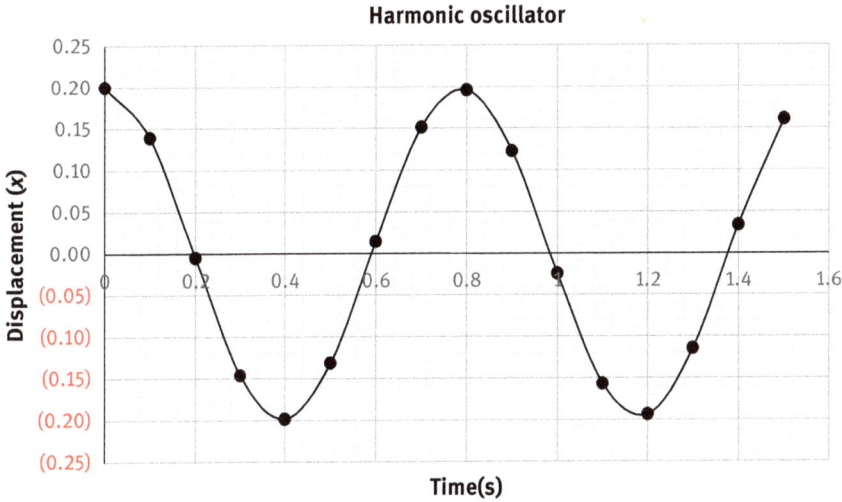

Figure 7.6: Harmonic motion of the spring solved by RK4 method.

7.4 Problems for practice

1. $\dfrac{dy}{dx} = y \ln \dfrac{y}{x}, \quad y(3) = e, h = 0.1$

2. $\dfrac{dy}{dx} = 30x^2 - \dfrac{8y}{3}, \quad y(0) = 0, h = 0.1$

3. $\dfrac{dy}{dx} = \sin(x + y) - e^x, \quad y(0) = 3, h = 0.1$

4. $\dfrac{d^2y}{dx^2} + \dfrac{dy}{dx} - 6y = 0 \quad \text{given } y(0) = 3 \text{ and } y'(0) = 1$

5. $\dfrac{d^2y}{dx^2} + 20\dfrac{dy}{dx} + 64y = 0 \quad y(0) = 0 \text{ and } y'(0) = 0.6$

Chapter 8
Numerical differentiation

8.1 Introduction

For a given point on the curve, the slope of the tangent to the curve gives the derivative of that function (curve). Numerical differentiation refers to calculating the derivative of an unknown function at some assigned point when the discrete data points are given. If we are given the set of data points for which the function is not given but only the data, then we can use those points to determine derivatives at those points. Numerical differentiation is the way of finding numerical value of derivative of a given function at a given point. In general, the derivative can be expressed as

$$f'(x) = \frac{f(x+h) - f(x)}{h} \qquad (8.1)$$

where h is the difference between two x values (e.g. x and $x + h$).

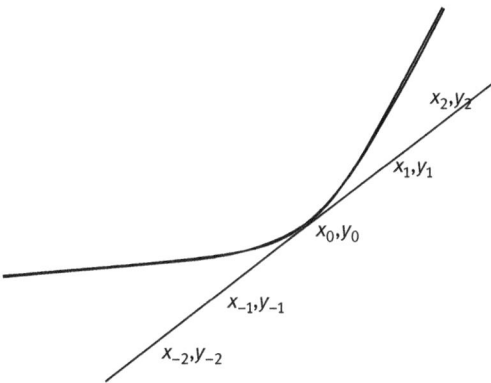

Figure 8.1: Tangent to the curve.

The derivative is the slope of the line that is made by connecting the data points (Figure 8.1). The derivative ($f'(x)$) at the point x_0 is

$$\frac{dy}{dx} = f'(x_0) = \frac{f(x_1) - f(x_0)}{x_1 - x_0} = \frac{y_1 - y_0}{x_1 - x_0} \qquad (8.2)$$

In general,

$$\frac{dy}{dx} = f'(x_i) = \frac{f(x_{i+1}) - f(x_i)}{x_{i+1} - x_i} = \frac{y_{i+1} - y_i}{x_{i+1} - x_i} \qquad (8.3)$$

https://doi.org/10.1515/9783111334448-008

This is also called forward difference formula (FDF). Similarly, if the two points are x_0, y_0 and x_{-1}, y_{-1}, that is backward points are considered, then

$$\frac{dy}{dx} = f'(x_0) = \frac{f(x_0) - f(x_{-1})}{x_0 - x_{-1}} = \frac{y_0 - y_{-1}}{x_0 - x_{-1}} \tag{8.4}$$

which can be written as

$$\frac{dy}{dx} = f'(x_i) = \frac{f(x_i) - f(x_{i-1})}{x_i - x_{i-1}} = \frac{y_i - y_{i-1}}{x_i - x_{i-1}} \tag{8.5}$$

This is known as backward difference formula (BDF). If a more balanced approach is taken to find derivative at x_0, take one backward point and one forward point as

$$\frac{dy}{dx} = f'(x_i) = \frac{f(x_{i+1}) - f(x_{i-1})}{x_{i+1} - x_{i-1}} = \frac{y_{i+1} - y_{i-1}}{x_{i+1} - x_{i-1}} \tag{8.6}$$

This is called central difference formula (CDF). This formula can also be derived using Taylor's series expansion. These FDF, BDF and CDF are the simplest approach to find derivative in case of a straight line case. But one does not have a straight line always, in such cases interpolating polynomial is used to find the derivative.

Derivation using Taylor's series

1. Approximation of a derivative in one variable
The first derivative from calculus can be simply written as

$$\frac{df}{dx} = f'(x) \approx \lim_{h \to 0} \frac{f(x+h) - f(x)}{h} \tag{A.1}$$

For a finite value of h, the approximation of the derivative can be written as

$$f'(x) \approx \frac{f(x+h) - f(x)}{h} + \Delta E \tag{A.2}$$

where ΔE is the error. This formula is the general "finite difference formula" for first derivative. If we want to estimate the error ΔE, we should consider the Taylor's series expansion of $x+h$ around the neighbourhood x as

$$f(x+h) = f(x) + hf'(x) + \frac{h^2}{2!}f''(x) + \frac{h^3}{3!}f'''(x) + \frac{h^4}{4!}f^{iv}(x) + \ldots \tag{A.3}$$

which can also be rewritten as

$$f'(x) = \frac{f(x+h) - f(x)}{h} - \frac{h^2}{2!}f''(x) - \frac{h^3}{3!}f'''(x) - \frac{h^4}{4!}f^{iv}(x) - \ldots \tag{A.4}$$

Equation (A.2) was obtained by truncating all the terms but first two terms in eq. (A.4). Since the approximation of the derivative of x above is based on the the values of x and x and $x+h$, this is called forward difference formula. Here ΔE as can be seen as a function of h, can be reduced to eq. (A.1) by truncating ΔE term, hence ΔE is also called truncation error. so

$$\Delta E = \frac{h}{2}f''(x) \tag{A.5}$$

Similarly if the Taylor's series expansion (eq. (A.3)) is written in the neighbourhood of $x-h$ then we have

$$f(x-h) = f(x) - hf'(x) + \frac{h^2}{2!}f''(x) - \frac{h^3}{3!}f'''(x) + \frac{h^4}{4!}f^{iv}(x) + \ldots \tag{A.6}$$

which can be rewritten as

$$f'(x) = \frac{f(x)-f(x-h)}{h} + \frac{h^2}{2!}f''(x) - \frac{h^3}{3!}f'''(x) + \frac{h^4}{4!}f^{iv}(x) + \ldots \tag{A.7}$$

Again retaining only first two terms we get,

$$f'(x) = \frac{f(x)-f(x-h)}{h} + \frac{h^2}{2!}f''(x) \tag{A.8}$$

which is of the form

$$f'(x) = \frac{f(x)-f(x-h)}{h} + \Delta E \tag{A.9}$$

Where ΔE is the truncation error.

$$f'(x) = \frac{f(x)-f(x-h)}{h} \tag{A.10}$$

Equation (A.10) is known as the BDF and ΔE is its truncation error which is same as in eq. (A.5). So far we have assumed that $f'(x)$ is continuous, now assume that $f'(x)$ is also continuous. A three term Taylor's series expansion of $f(x)$ about the point $x+h$ and $x-h$, respectively, is as follows

$$f(x+h) = f(x) + hf'(x) + \frac{h^2}{2!}f''(x) + \frac{h^3}{3!}f'''(x) \tag{A.11}$$

$$f(x-h) = f(x) - hf'(x) + \frac{h^2}{2!}f''(x) - \frac{h^3}{3!}f'''(x) \tag{A.12}$$

Subtracting (A.12) from (A.11)

$$f(x+h) - f(x-h) = 2hf'(x) + \frac{h^3}{3!}[f'''(\xi_2) + f'''(\xi_3)] \tag{A.13}$$

which on rearranging gives

$$f'(x) = \frac{f(x+h)-f(x-h)}{2h} - \frac{h^2}{12}[f'''(\xi_2) + f'''(\xi_3)] \tag{A.14}$$

So the truncation error in eq. (A.14) is

$$\Delta E = -\frac{h^2}{12}[f'''(\xi_2) + f'''(\xi_3)] \tag{A.15}$$

Also according to the intermediate theorem

$$f'''(\xi) = \frac{1}{2}[f'''(\xi_2) + f'''(\xi_3)] \tag{A.16}$$

So

$$\Delta E = -\frac{h^2}{6}f'''(\xi) \tag{A.17}$$

ΔE is the truncation error for two points CDF. Hence, eq. (A.14) becomes

$$f'(x) = \frac{f(x+h)-f(x-h)}{2h} - \frac{h^2}{6}f'''(\xi) \tag{A.18}$$

Hence, eq. (A.18) is a second-order approximation of first derivative. Likewise, other higher order approximations of derivative can be derived using Taylor's series expansion:

$$f'(x) = \frac{f(x+h)-f(x-h)}{2h} \tag{A.19}$$

Formula (A.19) is known as CDF.

Interpolating polynomials are the general expression of polynomials that are used to approximate the function $f(x)$ which is then used to find the derivative by differentiating the polynomial directly. There are many ways to estimate derivative using interpolation polynomial. Interpolation can thus also be used to estimate the value between a set of data. For this, first determine the appropriate interpolating polynomial for the given data. To carry out numerical differentiation, three types of interpolating polynomials, namely, Newton's interpolating polynomial, Stirling's polynomial and Lagrange's polynomial are used.

(a) When the data points are equidistant then
  (1) Newton's forward interpolation formula is used to find the derivative (near the beginning of the table).
  (2) Newton's backward interpolation polynomial is used to find the derivative (near the end of the table)
  (3) Stirling's formula is used to find the derivative (near the centre of the table).

(b) When the data points are not equidistant then
  (1) Newton's divided difference interpolation formula is used.
  (2) Lagrange's interpolation formula is used to find the derivative.

While if the data points are very far from each other, then Richardson extrapolation is used to find the derivative.

8.2 Derivatives using Newton's forward interpolation polynomial

Newton's forward interpolation formula is used to interpolate values of $[x,y(x)]$ near the beginning of the data set values $(x_i, y(x_i))$. Newton's forward interpolation formula (also called Newton–Gregory formula) is given by

$$y \equiv y_0 + p\Delta y_0 + \frac{p(p-1)}{2!}\Delta^2 y_0 + \frac{p(p-1)(p-2)}{3!}\Delta^3 y_0 + \frac{p(p-1)(p-2)(p-3)}{4!}\Delta^4 y_0 + \cdots \tag{8.7}$$

where $p = \frac{x - x_0}{h}$, y_0 is the first value of the data set. Δy_0 is the difference between two consecutive y values while $\Delta^2 y_0$ is the difference between two Δy_0 values and so on. To find the derivative, the function y is differentiated with respect to p:

$$\frac{dy}{dp} = \Delta y_0 + \frac{2p - 1}{2!} \Delta^2 y_0 + \frac{3p^2 - 6p + 2}{3!} \Delta^3 y_0 + \frac{4p^3 - 18p^2 + 22p - 6}{4!} \Delta^4 y_0 + \dots \qquad (8.8)$$

Also one may write

$$\frac{dy}{dx} = \frac{dy}{dp}\frac{dp}{dx} \qquad (8.9)$$

Since

$$\frac{dp}{dx} = \frac{1}{h}$$

$$\frac{dy}{dx} = \frac{1}{h}\left[\Delta y_0 + \frac{2p - 1}{2!} \Delta^2 y_0 + \frac{3p^2 - 6p + 2}{3!} \Delta^3 y_0 + \frac{4p^3 - 18p^2 + 22p - 6}{4!} \Delta^4 y_0 + \dots\right] \qquad (8.10)$$

At $x = x_0$, $p = 0$

$$\left.\frac{dy}{dx}\right]_{x=x_0} = \frac{1}{h}\left[\Delta y_0 - \frac{\Delta^2 y_0}{2!} + \frac{\Delta^3 y_0}{3!} - \frac{\Delta^4 y_0}{4!} + \dots\right] \qquad (8.11)$$

Similarly

$$\frac{d^2 y}{dx^2} = \frac{d}{dx}\left(\frac{dy}{dx}\right) = \frac{d}{dp}\left(\frac{dy}{dx}\right)\cdot\frac{dp}{dx} \qquad (8.12)$$

$$\left.\frac{d^2 y}{dx^2}\right]_{x=x_0} = \frac{1}{h^2}\left[\Delta^2 y_0 - \Delta^3 y_0 + \frac{11}{12}\Delta^4 y_0 - \frac{5}{6}\Delta^5 y_0 + \dots\right] \qquad (8.13)$$

For finding the maxima and minima of a function, $\frac{dy}{dx} = 0$.

$1/h \neq 0$, hence ignoring the higher order terms in eq. (8.10), a quadratic equation is obtained of the form

$$a + bp + cp^2 = 0 \qquad (8.14)$$

Solving for p gives the maxima/minima of the function $y = f(x)$ where $x = x_0 + ph$.

Example 1: Find the derivative **dy/dx** at $x = 2$ using the given data points.

x	0	1	2	3	4	5	6
y	-7	-2	5	14	25	38	53

Solution: Since the data points are equidistant (equally spaced x's value) and the derivative is to be found at the beginning of the table (at $x = 1$), Newton's forward interpolating polynomial is used. The given data points may be further tabulated in Table 8.1.

Table 8.1: Newton's forward difference interpolating formula.

x	y	Δy_0	$\Delta^2 y_0$	$\Delta^3 y_0$	$\Delta^4 y_0$
0	−7				
		5			
1	−2		2		
		7		0	
2	5		2		0
		9		0	
3	14		2		0
		11		0	
4	25		2		0
		13		0	
5	38		2		
		15			
6	53				

$$\left.\frac{dy}{dx}\right]_{x=x_0} = \frac{1}{h}\left[\Delta y_0 - \frac{\Delta^2 y_0}{2!} + \frac{\Delta^3 y_0}{3!} - \frac{\Delta^4 y_0}{4!} + \dots\right]$$

$$\left.\frac{dy}{dx}\right]_{x=1} = \frac{1}{1}\left[7 - \frac{2}{2!} + \frac{0}{3!}\right]$$

$$\left.\frac{dy}{dx}\right]_{x=1} = 6$$

One may also interpolate the given data and using curve fitting technique, and the general polynomial equation for this data is $y = x^2 + 4x - 7$ for which the derivative is $y' = 2x + 4$, hence, the solution.

Example 2: The following data of a first-order reaction was given as follows:

t/s	0	200	400	600	800	1000	1200
[A]	0.250	0.223	0.198	0.174	0.152	0.134	0.120

Find the slope **dy/dx** which gives rate constant at $t = 200$s.
Solution: For first order reactions, the integrated rate expression is given by

$$\ln\frac{[A_t]}{[A_0]} = -kt \tag{8.15}$$

Slope can give the rate constant when a graph between ln[A] vs t is plotted. Since the derivative is to be found out at the start of the observation table ($t = 200$s), so using the Newton forward Interpolation formula for derivative and constructing the Newton's forward difference construction table (Table 8.2).

Table 8.2: Newton's forward difference formula for kinetics data.

t = x	[A]	y = ln[A]	Δy	Δy²	Δy³	Δy⁴	Δy⁵	Δy⁶
0	0.25	−1.386						
			−0.114					
200	0.223	−1.500		−0.004				
			−0.118		−0.005			
400	0.198	−1.619		−0.010		0.010		
			−0.129		0.004		0.0007	
600	0.174	−1.748		−0.005		0.010		−0.0200
			−0.135		0.015		−0.0192	
800	0.152	−1.883		0.009		−0.008		
			−0.126		0.006			
1000	0.134	−2.009		0.015				
			−0.110					
1200	0.12	−2.120						

$$\left.\frac{dy}{dx}\right]_{t=200} = \frac{1}{200}\left[-0.118 - \frac{(-0.010)}{2} + \frac{0.004}{6} - \frac{0.010}{24} + \frac{(-0.01929)}{60}\right]$$

$$\left.\frac{dy}{dx}\right]_{t=200} = -0.00059$$

Hence, $\qquad k_{200} = 0.00059 s^{-1}$

Analytically solving from the given data, the rate constant is found to be $0.000650 s^{-1}$ which is almost the same as found by numerical differentiation.

There would always be some difference between the analytical answer and the answer obtained by numerical methods since they all are approximate methods.

8.3 Derivatives using Newton's backward interpolating polynomial

If the data point is to be estimated near the end of the data table, then Newton's backward interpolating polynomial is used which is

$$y = y_0 + p\Delta y_0 + \frac{p(p-1)}{2!}\Delta^2 y_0 + \frac{p(p-1)(p-2)}{3!}\Delta^3 y_0 + \frac{p(p-1)(p-2)(p-3)}{4!}\Delta^4 y_0 + \dots$$

(8.16)

$$\frac{dy}{dx} = \frac{dy}{dp}\frac{dp}{dx}$$

(8.9)

$$y = \frac{1}{h}\left[\Delta y_0 + \frac{2p+1}{2!}\Delta^2 y_0 + \frac{3p^2+6p+2}{3!}\Delta^3 y_0 + \frac{4p^3+18p^2+22p+6}{4!}\Delta^4 y_0 + \dots\right]$$

(8.17)

At $x = x_0$, $p = 0$

$$\left.\frac{dy}{dx}\right]_{x=x_0} = \frac{1}{h}\left[\Delta y_0 + \frac{\Delta^2 y_0}{2} + \frac{\Delta^3 y_0}{3} + \frac{\Delta^4 y_0}{4} + \dots\right] \tag{8.18}$$

$$\frac{d^2 y}{dx^2} = \frac{d}{dx}\left(\frac{dy}{dx}\right) = \frac{d}{dp}\left(\frac{dy}{dx}\right)\cdot\frac{dp}{dx} \tag{8.12}$$

Similarly

$$\left.\frac{d^2 y}{dx^2}\right]_{x=x_0} = \frac{1}{h^2}\left[\Delta^2 y_0 + \Delta^3 y_0 + \frac{11}{12}\Delta^4 y_0 + \frac{5}{6}\Delta^5 y_0 + \dots\right] \tag{8.19}$$

Example 3: Calculate the derivative dy/dx at $x = 1.5$ from the following data:

x	1	1.1	1.2	1.3	1.4	1.5	1.6
y	7.989	8.403	8.781	9.129	9.451	9.75	10.031

Solution: Since the derivative needs to be calculated at the end of the table, Newton's backward difference polynomial for calculating derivative is used. (Equation 8.18)

Table 8.3: Newton's backward difference interpolation formula for calculating derivative.

x	y	Δy₀	Δ²y₀	Δ³y₀	Δ⁴y₀	Δ⁵y₀	Δ⁶y₀
1	7.989						
		0.414					
1.1	8.403		−0.036				
		0.378		0.006			
1.2	8.781		−0.03		−0.002		
		0.348		0.004		0.001	
1.3	9.129		−0.026		0.001		0.002
		0.322		0.003		0.003	
1.4	9.451		0.023		0.002		
		0.299		0.005			
1.5	9.75		−0.018				
		0.281					
1.6	10.031						

$$\left.\frac{dy}{dx}\right]_{x=x_0} = \frac{1}{h}\left[\Delta y_0 + \frac{\Delta^2 y_0}{2!} + \frac{\Delta^3 y_0}{3!} + \frac{\Delta^4 y_0}{4!} + \dots\right]$$

$$\left.\frac{dy}{dx}\right]_{x=1.5} = \frac{1}{0.1}\left[0.299 + \frac{(-0.023)}{2} + \frac{0.03}{3} + \frac{(-0.001)}{4} + \frac{0.001}{5}\right]$$

$$\left.\frac{dy}{dx}\right]_{x=1.5} = 2.884$$

Example 4: Using Example 2, calculate the derivative at $t = 1,000$ s.

Solution: Since the derivative needs to be calculated at the end of Table 8.4, Newton's backward difference polynomial would be used.

Table 8.4: Newton's backward difference interpolation formula.

$t = x$	[A]	$y = \ln[A]$	Δy	Δy^2	Δy^3	Δy^4	Δy^5	Δy^6
0	0.25	-1.386						
			-0.114					
200	0.223	-1.500		-0.004				
			-0.118		-0.005			
400	0.198	-1.619		-0.010		0.010		
			-0.129		0.004		0.0097	
600	0.174	-1.748		-0.005		0.010		-0.0200
			-0.135		0.015		-0.0192	
800	0.152	-1.883		0.009		-0.008		
			-0.126		0.006			
1000	0.134	-2.009		0.015				
			-0.110					
1200	0.12	-2.120						

Using the Newton backward interpolating polynomial as

$$\left.\frac{dy}{dx}\right]_{t=1,000} = \frac{1}{200}\left[-0.126 + \frac{0.009}{2} + \frac{0.015}{3} + \frac{0.010}{4} + \frac{(0.0007)}{5}\right]$$

$$\left.\frac{dy}{dx}\right]_{t=1,000} = -0.0005680$$

$$k_{1,000} = 0.0005680\ \text{s}^{-1}$$

8.4 Derivatives using Stirling interpolating polynomial

The Stirling formula computes the average of values obtained by Gauss forward and backward interpolation. The condition for using Stirling's formula is $-1/2 < p < 1/2$. The formula is given as

$$y \equiv y_0 + p\left(\frac{\Delta y_0 + \Delta y_{-1}}{2}\right) + \frac{p^2}{2!}\Delta^2 y_{-1} + \frac{p(p^2-1)}{3!}\left(\frac{\Delta^3 y_{-1} + \Delta^3 y_{-2}}{2}\right)$$

$$+ \frac{p^2(p^2-1)}{4!}\Delta^4 y_{-2} + \frac{p(p^2-1)(p^2-2^2)}{5!}\left(\frac{\Delta^5 y_{-3} + \Delta^5 y_{-2}}{2}\right) + \cdots$$

(8.20)

where $p = \dfrac{x - x_0}{h}$

Differentiating the above equation with respect to p,

$$\frac{dy}{dp} = \left(\frac{\Delta y_0 + \Delta y_{-1}}{2}\right) + p\Delta^2 y_{-1} + \frac{3p^2 - 1}{3!}\left(\frac{\Delta^3 y_{-1} + \Delta^3 y_{-2}}{2}\right) + \frac{4p^3 - 2p}{4!}\Delta^4 y_{-2} + \dots \quad (8.21)$$

Also $\dfrac{dp}{dx} = \dfrac{1}{h}$, so

$$\frac{dy}{dx} = \frac{dy}{dp}\cdot\frac{dp}{dx} \quad (8.9)$$

Therefore, we get

$$\frac{dy}{dx} = \frac{1}{h}\left[\left(\frac{\Delta y_0 + \Delta y_{-1}}{2}\right) + p\Delta^2 y_{-1} + \frac{3p^2 - 1}{3!}\left(\frac{\Delta^3 y_{-1} + \Delta^3 y_{-2}}{2}\right) + \frac{4p^3 - 2p}{4!}\Delta^4 y_{-2} + \dots\right] \quad (8.22)$$

At $x = x_0$, $p = 0$

$$\left.\frac{dy}{dx}\right]_{x=x_0} = \frac{1}{h}\left[\left(\frac{\Delta y_0 + \Delta y_{-1}}{2}\right) - \frac{1}{6}\left(\frac{\Delta^3 y_{-1} + \Delta^3 y_{-2}}{2}\right) + \frac{1}{30}\left(\frac{\Delta^5 y_{-2} + \Delta^5 y_{-3}}{2}\right) - \dots\right] \quad (8.23)$$

Similarly

$$\left.\frac{d^2 y}{dx^2}\right]_{x=x_0} = \frac{1}{h^2}\left[\Delta^2 y_{-1} - \frac{1}{12}\Delta^4 y_{-2} + \frac{1}{90}\Delta^6 y_{-3} - \dots\right] \quad (8.24)$$

Example 5: Find the derivative dy/dx at $x = 0.3$.
Solution: Since here in this example, the derivative needs to be found at the mid of the data table, Stirling's interpolation polynomial (Table 8.5) will be used (eq. (8.23)).

x	0	0.1	0.2	0.3	0.4	0.5	0.6
y	4	4.044	4.016	3.916	3.744	3.5	3.184

Table 8.5: Stirling's interpolation formula to find derivative.

x	Y	$\Delta^1 y_0$	$\Delta^2 y_0$	$\Delta^3 y_0$	$\Delta^4 y_0$	$\Delta^5 y_0$	$\Delta^6 y_0$
0	4						
		0.044					
0.1	4.044		−0.072				
		−0.028		−1.332E−15			
0.2	4.016		−0.072		1.776E−15		
		−0.1		4.44E−16		−1.77E−15	
0.3	3.916		−0.072		0		8.88E−16
		−0.172		4.44E−16		−8.88E−16	
0.4	3.744		−0.072		−8.88E−16		

Table 8.5 (continued)

x	Y	$\Delta^1 y_0$	$\Delta^2 y_0$	$\Delta^3 y_0$	$\Delta^4 y_0$	$\Delta^5 y_0$	$\Delta^6 y_0$
		−0.244		−4.44E−16			
0.5	3.5		−0.072				
		−0.316					
0.6	3.184						

$$\frac{dy}{dx} = \frac{1}{h}\left[\left(\frac{\Delta y_0 + \Delta y_{-1}}{2}\right) - \frac{1}{6}\left(\frac{\Delta^3 y_{-1} + \Delta^3 y_{-2}}{2}\right) + \frac{1}{30}\left(\frac{\Delta^5 y_{-2} + \Delta^5 y_{-3}}{2}\right) - \ldots\right]$$

$$\frac{dy}{dx} = \frac{1}{h}\left[\left(\frac{\Delta y_0 + \Delta y_{-1}}{2}\right) - \frac{1}{6}\left(\frac{\Delta^3 y_{-1} + \Delta^3 y_{-2}}{2}\right) + \frac{1}{30}\left(\frac{\Delta^5 y_{-2} + \Delta^5 y_{-3}}{2}\right) - \ldots\right]$$

$$\left.\frac{dy}{dx}\right]_{x=0.3} = \frac{1}{0.1}\left[\left(\frac{-0.1 + (-0.172)}{2}\right) - \frac{1}{6}\left(\frac{8.8817 \times 10^{-16} + 8.8817 \times 10^{-16}}{2}\right)\right.$$
$$\left. + \frac{1}{30}\left(\frac{-3.9968 \times 10^{-15} + (3.5528.88 \times 10^{-16})}{2}\right)\right]$$

$$\left.\frac{dy}{dx}\right]_{x=0.3} = -1.36$$

Example 6: For the reversible reaction

$$\frac{A_{eq} - A}{A_{eq}} = \frac{70 - (100\% - A\%)}{70}$$

The rate of change of A is as follows:

Time (h)	0	1	2	3	4	∞
% A	100	72.5	56.8	45.6	39.5	30

where the reaction follows the first-order kinetics with $\ln\left(\frac{A_{eq} - A}{A_{eq}}\right) = -(k_1 + k_{-1})t$ where A_{eq} and A are the equilibrium concentration and concentration at any time t. Calculate the rate constants for both forward and backward reaction and the equilibrium constant.

Solution: Since in this example, the data is given for concentration of A and we need log term as $\ln\left(\frac{A_{eq} - A}{A_{eq}}\right)$ so before applying the method of numerical differentiation in terms are calculated as in Table 8.6.

Table 8.6: Kinetics data for reversible reaction.

Time (h)	0	1	2	3	4	∞
% A	100	72.5	56.8	45.6	39.5	30
$\ln\left(\frac{A_{eq} - A}{A_{eq}}\right)$	0	−0.4992	−0.9597	−1.5005	−1.9951	−

As no particular data point is given to calculate the derivative, it is calculated at the middle interval. Calculating the derivative at the interval of 2 h using Stirling's polynomial method of derivative (Table 8.7), we may depict the same as $h = 1$ h:

$$\left.\frac{dy}{dx}\right]_{t=2h} = \frac{1}{h}\left[\left(\frac{\Delta y_0 + \Delta y_{-1}}{2}\right) - \frac{1}{6}\left(\frac{\Delta^3 y_{-1} + \Delta^3 y_{-2}}{2}\right) + \cdots\right]$$

$$\left.\frac{dy}{dx}\right]_{t=2h} = \frac{1}{1}\left[\left(\frac{-0.4605 + (-0.5409)}{2}\right) - \frac{1}{6}\left(\frac{-0.1191 + 0.1267}{2}\right)\right]$$

$$\left.\frac{dy}{dx}\right]_{t=2h} = -0.5013$$

Table 8.7: Stirling's interpolation polynomial for kinetics data.

x_i	y_i	Δy	$\Delta^2 y$	$\Delta^3 y$	$\Delta^4 y$
0	0.0000				
		-0.4992			
1	-0.4992		0.0387		
		-0.4605		-0.1191	
2	-0.9597		-0.0804		0.2458
		-0.5409		0.1267	
3	-1.5006		0.0463		
		-0.4945			
4	-1.9951				

where the slope graphically is found to be –0.4991.

Also $K = \frac{k_1}{k_{-1}} = \frac{[B]_{eq}}{[A]_{eq}} = \frac{70}{30} = 2.333$

Hence substituting the known values in the relationship, we get

$$k_1 = 0.3506\text{ h}^{-1}, \quad k_1 = 0.1502\text{ h}^{-1}$$

Higher order derivatives can be found out by subsequently differentiating the previous derivatives. Numerical differentiation using Stirling's polynomial is more accurate as compared to Newton's polynomial. When the points are usually unequally spaced, Newton's divided difference polynomial and Lagrange's polynomial is used.

8.5 Newton's divided difference polynomial

When the intervals are unequispaced, then Newton's divided difference interpolation formula is used which is

$$f(x) = f(x_0) + (x - x_0)f[x_0, x_1] + (x - x_0)(x - x_1)f[x_0, x_1, x_2]$$
$$+ (x - x_0)(x - x_1)(x - x_2)f[x_0, x_1, x_2, x_3] + \cdots$$

$$(8.25)$$

where $f[x_0,x_1] = \dfrac{y_1-y_0}{x_1-x_0}$

similarly $f[x_0,x_1,x_2] = \dfrac{\Delta y_2 - \Delta y_0}{x_2-x_0}$

The earlier equation is then differentiated with respect to x to find the derivative at the particular point. Apart from Newton's divided difference polynomial, Lagrange's polynomial is also used for interpolation of data points which are at unequal intervals.

Example 7: Using the given data, evaluate **dy/dx** at $x = 6$.

X	0	2	3	4	7	9
Y	4	26	58	112	466	922

Solution: Since data points are not equally spaced, the table for Newton's divided difference table (Table 8.8) for interpolation is constructed.

Table 8.8: Newton's divided difference formula for interpolation.

x	y	$f[x_0,x_1]$	$f[x_0,x_1,x_2]$	$f[x_0,x_1,x_2,x_3]$	$f[x_0,x_1,x_2,x_3,x_4]$
0	4				
		$\dfrac{26-4}{2-0}=11$			
2	26		$\dfrac{32-11}{3-0}=7$		
		$\dfrac{58-26}{3-2}=32$		$\dfrac{11-7}{4-0}=1$	
3	58		$\dfrac{54-32}{4-2}=11$		0
		$\dfrac{112-58}{4-3}=54$		$\dfrac{16-11}{7-2}=1$	
4	112		$\dfrac{118-54}{7-3}=16$		0
		$\dfrac{466-112}{7-4}=118$		$\dfrac{22-16}{9-3}=1$	
7	466		$\dfrac{278-118}{9-4}=22$		
		$\dfrac{922-466}{9-7}=278$			
9	922				

Using eq. (8.25)

$$f(x) = 4 + (x-0)11 + x(x-2)7 + x(x-2)(x-3)1$$
$$f(x) = x^3 + 2x^2 + 3x + 4$$
$$f'(x) = 3x^2 + 4x + 3$$
$$f'(x)_{x=6} = 108 + 24 + 3 = 135$$

8.6 Lagrange's polynomial

For the given data points x_1, x_2, x_3, . . ., x_n where they are not at equidistant, the Lagrange's interpolating polynomial for the points (x_1, y_1), (x_2, y_2), (x_3, y_3),. . ., (x_n, y_n) is given by

$$L(x) = L(x_1)f(x_1) + L(x_2)f(x_2) + L(x_3)f(x_3) \qquad (8.26)$$

$$L(x) = \frac{(x - x_2)(x - x_3)}{(x_1 - x_2)(x_1 - x_3)}f(x_1) + \frac{(x - x_1)(x - x_3)}{(x_2 - x_1)(x_2 - x_3)}f(x_2) + \frac{(x - x_1)(x - x_2)}{(x_3 - x_1)(x_3 - x_2)}f(x_3) \qquad (8.27)$$

The derivative of the function is the derivative of the Lagrange's polynomial itself; hence, we can write

$$f'(x) = L'(x) = L'(x_1)f(x_1) + L'(x_2)f(x_2) + L'(x_3)f(x_3) \qquad (8.28)$$

$$f'(x) \approx \frac{2x - x_2 - x_3}{(x_1 - x_2)(x_1 - x_3)}f(x_1) + \frac{2x - x_1 - x_3}{(x_2 - x_1)(x_2 - x_3)}f(x_2) + \frac{2x - x_1 - x_2}{(x_3 - x_1)(x_3 - x_2)}f(x_3) \qquad (8.29)$$

So if we want to find derivative at x_1 then

$$f'(x_1) \approx \frac{2x_1 - x_2 - x_3}{(x_1 - x_2)(x_1 - x_3)}f(x_1) + \frac{x_1 - x_3}{(x_2 - x_1)(x_2 - x_3)}f(x_2) + \frac{x_1 - x_2}{(x_3 - x_1)(x_3 - x_2)}f(x_3) \qquad (8.30)$$

If we consider the forward difference points such that $x_1 = x$, $x_2 = x + h$ and $x_2 = x + 2h$:

$$f'(x) \approx \frac{-3}{2h}f(x_1) + \frac{4}{2h}f(x_2) - \frac{1}{2h}f(x_3) \qquad (8.31)$$

which can also be written as

$$f'(x_1) \approx \frac{-3f(x) + 4f(x + h) - f(x + 2h)}{2h} \qquad (8.32)$$

This is known as three-point FDF.

The second derivative can also be written by differentiating the equation again and we get the general formula as

$$f''(x_i) \approx \frac{1}{h^2}[f(x_{i+1}) - 2f(x_i) + f(x_{i-1})] \qquad (8.33)$$

The above is the three-point FDF to find derivative in i^{th} position. If we take the four points x_0, x_1, x_2, x_3, then similarly the second derivative can also be found out as

$$f''(x) \approx L''(x) = \frac{2[(x - x_1) + (x - x_2) + (x - x_3)]}{(x_0 - x_1)(x_0 - x_2)(x_0 - x_3)}f(x_0) + \frac{2[(x - x_0) + (x - x_2) + (x - x_3)]}{(x_1 - x_0)(x_1 - x_2)(x_1 - x_3)}f(x_1)$$

$$+ \frac{2[(x - x_0) + (x - x_1) + (x - x_3)]}{(x_2 - x_0)(x_2 - x_1)(x_2 - x_3)}f(x_2) + \frac{2[(x - x_0) + (x - x_1) + (x - x_2)]}{(x_3 - x_0)(x_3 - x_1)(x_3 - x_2)}f(x_3) \qquad (8.34)$$

Considering the forward differences as

$x_0 = x$, $x_1 = x + h$, $x_2 = x + 2h$ and $x_3 = x + 3h$ we get

$$f''(x) \approx \frac{2f(x) - 5f(x+h) + 4f(x+2h) - f(x+3h)}{h^2} \tag{8.35}$$

The earlier formula is referred to as four-point FDF for second derivative.

Example 8: Using the given data, find the derivative using Lagrange's polynomial (Table 8.9).

Table 8.9: Lagrange's polynomial for interpolation for unequispaced data points.

N	0	1	2	3
x_n	−1	1	4	7
$y_n = f(x_n)$	−2	0	63	342

Solution: Lagrange's polynomial for interpolation can be written as

$$P_2(x) = L_0(x)f(x_0) + L_1(x)f(x_1) + L_2(x)f(x_2) + L_3(x)f(x_3) \tag{8.36}$$

where

$$L_0(x) = \frac{(x-x_1)(x-x_2)(x-x_3)}{(x_0-x_1)(x_0-x_2)(x_0-x_3)} = \frac{-1}{80}(x^3 - 12x^2 + 39x - 28)$$

$$L_1(x) = \frac{(x-x_0)(x-x_2)(x-x_3)}{(x_1-x_0)(x_1-x_2)(x_0-x_3)} = \frac{1}{36}(x^3 - 10x^2 + 17x + 28)$$

$$L_2(x) = \frac{(x-x_0)(x-x_1)(x-x_3)}{(x_2-x_0)(x_2-x_1)(x_2-x_3)} = \frac{-1}{45}(x^3 - 7x^2 - x + 7)$$

$$L_3(x) = \frac{(x-x_0)(x-x_1)(x-x_2)}{(x_3-x_0)(x_3-x_1)(x_3-x_2)} = \frac{1}{144}(x^3 - 4x^2 - x + 4)$$

Substituting the values of (x_n, y_n), one may write

$$P_3(x) = \frac{-1}{80}(x^3 - 12x^2 + 39x - 28)(-2) + \frac{1}{36}(x^3 - 10x^2 + 17x + 28)(0) +$$

$$\frac{-1}{45}(x^3 - 7x^2 - x + 7)(63) + \frac{1}{144}(x^3 - 4x^2 - x + 4)(342) \tag{8.37}$$

$$P_3(x) = x^3 - 1 \tag{8.38}$$

$$P_3'(x) = 3x^2 \tag{8.39}$$

$$P_3(5) = 75$$

Example 9: Find the derivative dP/dV using the given data for van der Waal's gas.

	x_0	x_1	x_2	x_3
x	1	2	4	5
$y = f(x)$	2224.242	1174.851	603.336	485.229

Solution: Here the points are not equidistant; hence, the derivative can be found out using either Newton divided difference formula or Lagrange's polynomial. Hence, using the Newton divided difference formula as (Table 8.10)

$$f(x) = f(x_0) + (x-x_0)f[x_0,x_1] + (x-x_0)(x-x_1)f[x_0,x_1,x_2] + (x-x_0)(x-x_1)(x-x_2)f[x_0,x_1,x_2,x_3] + ...$$

Table 8.10: Newton's divided difference interpolation polynomial data.

xi	yi	Δy	Δ²y	Δ³y
1	2224.242047			
		−1049.390382		
2	1174.851665		254.5442863	
		−285.757523		−49.66523638
4	603.336619		55.8833408	
		−118.1075006		
5	485.2291184			

Substituting the requisite values

$$P = f(V) = f(V_0) + (V-V_0)f[V_0,V_1] + (V-V_0)(V-V_1)f[V_0,V_1,V_2]$$
$$P = 2224.24 - (V-1)1049.39 + (V-1)(V-2)254.54 - (V-1)(V-2)(V-3)49.66$$

and differentiating with respect to V by subsuming $V = 1$, one gets

$$dP/dV = -1452.9303$$

It means that pressure decreases with increase in volume which is nothing but Boyle's law.
    Similarly using the Lagrange's polynomial for interpolation for unequal spaces as

$$L(x) = \frac{(x-x_1)(x-x_2)}{(x_0-x_1)(x_0-x_2)}f(x_0) + \frac{(x-x_2)(x-x_3)}{(x_1-x_2)(x_1-x_3)}f(x_1) + \frac{(x-x_1)(x-x_3)}{(x_2-x_1)(x_2-x_3)}f(x_2) + \frac{(x-x_1)(x-x_2)}{(x_3-x_1)(x_3-x_2)}f(x_3)$$

Here using x_0, x_1, x_2 and $x_3 = 1, 2, 4$ and 5, respectively, we get

$$l_0(x) = \frac{(x-2)(x-4)(x-5)}{-12}, \quad l_1(x) = \frac{(x-1)(x-4)(x-5)}{6}, \quad l_2(x) = \frac{(x-1)(x-2)(x-5)}{-6}$$

$$l_3(x) = \frac{(x-1)(x-2)(x-4)}{12}$$

Substituting back in earlier formula for $L(x)$ and differentiating with respect to x and substituting the values of the requisite function values at $x = 1$, we get −1452.9303.

Both the methods give the same answer. When solved analytically using the Van der Waal's equation derivative, the answer is −1975.83. Being approximate methods, again the answer obtained numerically will most of the times differ from conventional answer.

8.7 Richardson's extrapolation

Sometimes when we use the data points that are not very close (Figure 8.2), for example, x_i and x_{i+1} and try to find the derivative or slope we get large amount of error. It is so because the difference between the interval (h) of x is very large.

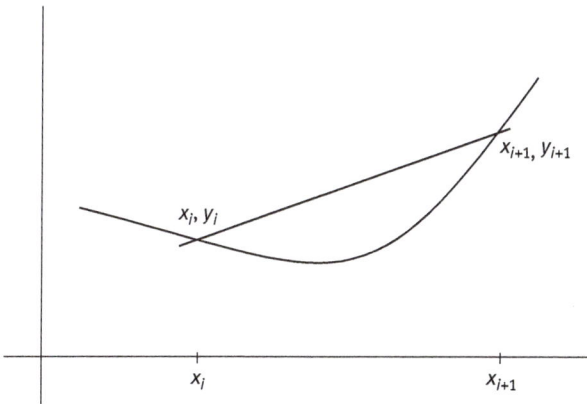

Figure 8.2: Richardson's extrapolation when data points are very far from each other.

If we reduce the h value, the slope value will improve like Figure 8.3.

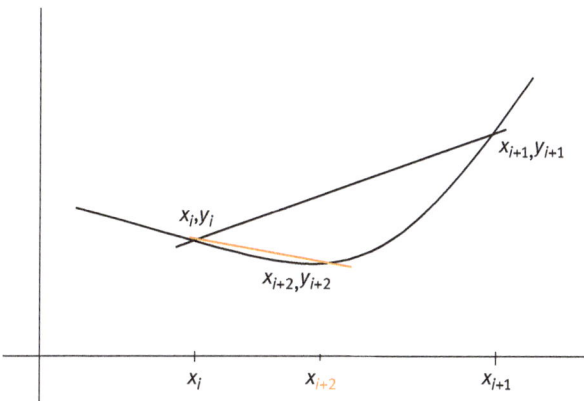

Figure 8.3: Decreasing the interval size.

Now we can see that when the interval was reduced the slope of the line improves and one gets less error. Hence one can say that the accuracy of the method for estimating the derivative or integral of the function $f(x)$ depends on the spacing between points at which the integral is evaluated. The approximation tends to improve as the spacing tends to zero.

With h being the uniform spacing, the data points may be approximated by the expression as

$$D(h) = f'(x_i) = \left[\frac{f(x_{i+h}) - f(x_{i-h})}{2h}\right] - \frac{h^2}{6}f'''(x) + O(h^4) \tag{8.40}$$

$D(h)$ is the approximated value by using CDF with step size h. Similarly we can write the formula when the interval is $h/2$ as

$$D(h/2) = f'(x_i) = \left[\frac{f(x_{i+h/2}) - f(x_{i-h/2})}{2(h/2)}\right] - \frac{h^2}{64}f'''(x) + O(h^4) \tag{8.41}$$

$D(h/2)$ is the approximated value by using CDF with step size $h/2$, solving the earlier two equations we get

$$D = \frac{4D(h/2) - D(h)}{3} \tag{8.42}$$

D is the improved estimate or true value. This process of extrapolation from $D(h)$ and $D(h/2)$ to approximate D with a higher order accuracy is called Richardson's extrapolation. The earlier formula explains the computation of approximation that is fourth-order accurate. This method is further extended to achieve greater accuracy up to order of $O(h^6)$, $O(h^8)$ and so on. The general formula for orders $O(h^6)$ or higher than that is

$$D^k(h) = \frac{4^k D^{k-1}(h/2) - D^{k-1}(h)}{4^k - 1} \tag{8.43}$$

Example 10: Using the following data (Table 8.11), calculate derivative $f'(x_2)$ where $h = 2$.

Table 8.11: Richardson's extrapolation.

n	0	1	2	3	4
x_n	1	2	3	4	5
y_n	2	4	8	16	32

Solution: Using Richardson extrapolation,

$$D(h) = f'(x_2) = \frac{1}{2h}[f(x_2 + h) - f(x_2 - h)]$$

$$D(h) = f'(x_2) = \frac{1}{4}[f(5) - f(1)]$$

$$D(h) = f'(x_2) = \frac{1}{4}[32 - 2] = 7.5$$

Similarly calculating $D(h/2)$

$$D(h/2) = f'(x_2) = \frac{1}{2\frac{h}{2}}\left[f\left(x_2 + \frac{h}{2}\right) - f\left(x_2 - \frac{h}{2}\right)\right]$$

$$D(h/2) = f'(x_2) = \frac{1}{2}[f(4) - f(2)]$$

$$D(h/2) = f'(x_2) = \frac{1}{2}[16 - 4] = 6$$

Hence using eq. (8.41),

$$D = \frac{4D(h/2) - D(h)}{3}$$

$$D = \frac{4 \times 6 - 7.5}{3} = \frac{16.5}{3} \approx 5.5$$

The actual function is $y = 2^x$, whose derivative $y' = 2^x \ln|2| \approx 5.5443$.

This formula was to obtain an accuracy of fourth order. One may also achieve accuracy of higher order.

8.8 Problems for practice

1. Find the derivative $\frac{dy}{dx}$ and $\frac{d^2y}{dx^2}$ at $x = 5$ using the given data points.

x	1	2	3	4	5	6
y	−2	5	14	25	38	53

2. Using the Stirling's polynomial for interpolating derivatives, find the most probable speed in Maxwell's distribution curve here:

$$\frac{1}{N}\frac{dN}{du} = \int\limits_{0}^{1,200} 4\pi u^2 \left(\frac{m}{2\pi RT}\right)^{3/2} \exp\left(\frac{-Mu^2}{2RT}\right)$$

u	0	200	400	600	800	1,000	1,200	1,400
$\frac{1}{N}\frac{dN}{du} = y$	0	0.001144	0.002109	0.001304	0.000380	0.000058	0.000005	0

3. Using the given data, find x where y is maximum.

x	3	4	5	6	7	8
y	0.205	0.240	0.259	0.262	0.250	0.224

4. Find dy/dx at $x = 1$ using the following data:

x	0	2	4	6	8
y	4	8	15	7	6

Chapter 9
Numerical root-finding methods

9.1 Introduction

In chemistry, we often come across lengthy and complicated polynomial equations, which are difficult to solve analytically. According to algebra, a root is the zero of the function, that is, where the function $f(x)$ is zero. There are three ways to solve the equations, namely analytically, graphically and numerically. Numerical methods of finding roots of the equations is the most robust way of solving even very complicated equation with a great degree of ease. The most important technique in any numerical method is the iteration. Generally, an approximation of an expected value is taken and an algorithm is applied which further improves the approximation. This step is repeated until the approximation yields almost the same value. Numerical methods are particularly useful while solving the intensive polynomial for their roots.

These are the following numerical methods to find roots of an equation:
1. Newton–Raphson method
2. Iteration method
3. Binary bisection method
4. Secant method
5. Regula-Falsi method

Each of the above-mentioned methods is discussed at length in the subsequent sections with their drawbacks and advantages.

9.2 Newton–Raphson method

Newton–Raphson (also called Newton's iteration or Newton's technique) is the most widely used root-finding algorithm of nonlinear equations or real-valued single variable functions ($f(x) = 0$). It uses an iterative method to approach the root of equation by arbitrarily choosing any root which is close to the real root.

N-R method converges quadratically as we approach the root. It needs only one initial guess value for the root. This method involves expansion of Taylor series of a function $f(x)$ in the vicinity of suspected root and only first few terms (usually up to second term) of the Taylor series are retained to find the real root. Hence, truncation at second term gives the estimate of the root.

https://doi.org/10.1515/9783111334448-009

For N-R method, function $f(x)$ can be written using a Taylor's expansion as

$$f(x_0 + h) = f(x_0) + hf'(x_0) + \frac{h^2}{2!}f''(x_0) + \cdots = 0 \tag{9.1}$$

where $f'(x_0)$ and $f''(x_0)$ are the first and second derivatives of $f(x_0)$ with respect to x and $h = x - x_0$. Since $h \lll 1$, h^2 can be neglected; hence, one may write

$$f(x_0) + hf'(x_0) = 0 \tag{9.2}$$

which gives

$$h = -\frac{f(x_0)}{f'(x_0)} \tag{9.3}$$

Therefore,

$$x = x_0 - \frac{f(x_0)}{f'(x_0)} \tag{9.4}$$

Hence, the successive approximations for Newton–Raphson method can be written as a general formula

$$x_{n+1} = x_n - \frac{f(x_n)}{f'(x_n)} \tag{9.5}$$

This equation is nothing but the equation of the tangent to the function $f(x)$ at $x = x_n$ which is extrapolated to x-axis to get x_{n+1} (first approximate root). Consider the function $f(x)$ which is continuous such that $f'(x) \neq 0$.

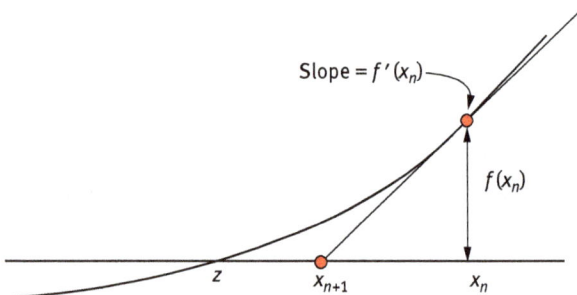

Figure 9.1: Tangent to the curve to approximate the root.

This method assumes that functions that have continuous derivatives and look like straight lines when looked closely can be approximated by a tangent to it. Assuming $f(x)$ has a continuous derivative, $f'(x)$ can be computed. Since $f'(x) \neq 0$, it implies that any point on the $f(x)$ curve will be a straight line. The function at $(x_n, f(x_n))$ can be approximated by a straight line that is tangent to the curve at that point (Figure 9.1). The slope of the line is $f'(x)$ which passes through $(x_n, f(x_n))$. Therefore, the equation of the tangent line is

$$y - y_1 = m(x - x_1) \tag{9.6}$$

$$y - f(x_n) = f'(x_n)(x - x_n) \tag{9.7}$$

The tangent crosses the x-axis at $x = x_{n+1}$, at which $y = 0$; hence,

$$0 - f(x_n) = f'(x_n)(x_{n+1} - x_n) \tag{9.8}$$

which may be rearranged as

$$x_{n+1} = x_n - \frac{f(x_n)}{f'(x_n)} \tag{9.5}$$

which is same as obtained earlier. Hence, the first approximation x_n has improved to x_{n+1}, which is much nearer to the real root z. This approximation may then be used to find another approximation until the real root (z) is reached. N-R greatly depends upon the initial guess value, so the starting guess value should be chosen carefully such that x_0 is close to the original root z.

When the difference between the successive iterates is small enough to follow $|x_{n+1} - x_n| < \varepsilon_{abs}$, then x_{n+1} is the approximate root of the equation. Also if $f(x_{n+1})$ is sufficiently small enough such that $|f(x_{n+1})| \approx 0$, then x_{n+1} is the approximation to the root.

Example 1: Using N-R method, find the approximate root for the following function $f(x)$:

$$f(x) = x^2 + 5x - 3 = 0 \tag{9.9}$$

Solution: The above function can be solved graphically (Figure 9.2) as well as numerically. The roots are those which cut the x-axis, where $f(x) = 0$. The order of polynomial refers to the number of roots the function has, which is two here. Graphically, it may be pointed that root lies between [0,1] and another root between [−5,−6]. For the above function $f(x)$, its derivative $f'(x) = 2x + 5$.

We can find both roots by employing N-R method using two different guesses. Taking the initial guess $x_0 = 2$, the results are tabulated in Table 9.1.

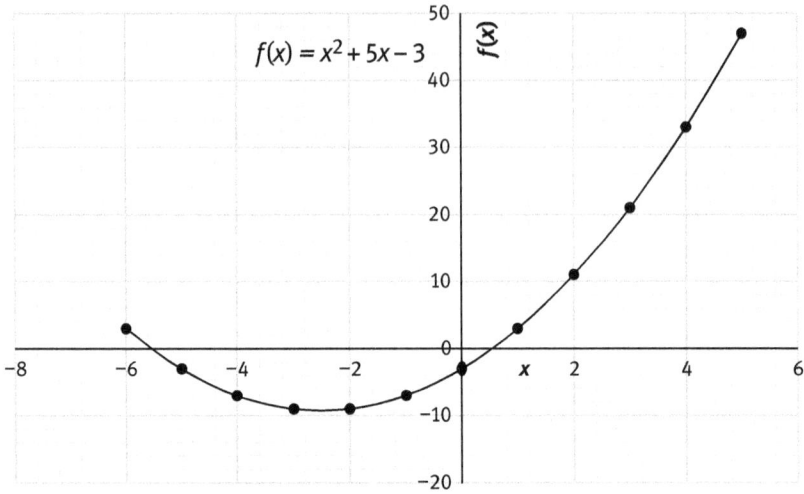

Figure 9.2: Solving $f(x)$ graphically.

Table 9.1: N-R method to solve $f(x) = x^2 + 5x-3$.

N	x_n	x_{n+1}	$f(x)$	$f'(x)$	ε_{abs}
1	2	0.777777778	11	9	1.222222222
2	0.777777778	0.549905838	1.49382716	6.555555556	0.22787194
3	0.549905838	0.541393178	0.051925621	6.099811676	0.00851266
4	0.541393178	0.541381265	7.24654E-05	6.082786357	1.1913E-05
5	0.541381265	0.541381265	1.41924E-10	6.08276253	0

As evident from the table, $f(x)$ becomes zero at the fifth iteration. Also at fifth iteration, difference between $(x_{n+1} -x_n) = 0$, so one of the roots is 0.5413812651. The difference between two roots is less than 0.000001(ε_{abs}) or even becomes zero on successive iterations. Similarly, when another guess is taken which is close enough to another root, the another root is found to be −5.541381265 (Table 9.2).

Table 9.2: N-R method to find root using another guess value for $f(x) = x^2 + 5x-3$.

N	x_n	x_{n+1}	$f(x)$	$f'(x)$	ε_{abs}
1	−3	−12	−9	−1	9
2	−12	−7.736842105	81	−19	4.263157895
3	−7.736842105	−6.001586882	18.17451524	−10.47368421	1.735255223
4	−6.001586882	−5.571623155	3.011110691	−7.003173764	0.429963727

Table 9.2 (continued)

N	x_n	x_{n+1}	f(x)	f'(x)	ε_{abs}
5	-5.571623155	-5.54153014	0.184868806	-6.14324631	0.030093015
6	-5.54153014	-5.541381269	0.00090559	-6.083060279	0.000148871
7	-5.541381269	-5.541381265	2.21625E-08	-6.082762538	3.64349E-09

Although N-R is a very good method for approximating the roots, there are some limitations. N-R method may fail if

(1) The approximation initial value (guess) is very far from the real root.
(2) The derivative of function is bit complicated. For example

$$f(x) = \sin(\cos(e^x)) \text{ and } f''((x) = -e^x \sin(e^x)\cos(\cos(e^x))$$

In such cases, secant method serves as a saviour where instead of a tangent line a secant is used.

(3) It may also not converge if the second derivative is very large or if the derivative at x_n is near to zero.
(4) There is local maxima or minima or point of inflection around the root.

Example 2: Find an approximation of $\sqrt{7}$ to 10 decimal places.

Solution: $\sqrt{7}$ is an irrational number, hence, sequence of decimals will be non-stop. $x = \sqrt{7}$ is the zero of the equation $f(x) = x^2 - 7$ on the interval [2,3].
So, here, using the N-R method having n approximation, the formula used is

$$x = x_0 - \frac{f(x_0)}{f'(x_0)}$$

With the initial guess value be $x_0 = 2$, the approximation is tabulated (Table 9.3).

Table 9.3: N-R method for to approximate.

N	x_n	x_{n+1}	f(x)	f'(x)
1	2	2.75	-3	4
2	2.75	2.647727273	0.5625	5.5
3	2.647727273	2.645752048	0.010459711	5.295454545
4	2.645752048	2.645751311	3.90151E-06	5.291504097
5	2.645751311	2.645751311	5.43565E-13	5.291502622

The correct approximation for $\sqrt{7}$ when deduced analytically is also found to be same.

Example 3: The inversion temperature of van der Waal's gas can be calculated using the equation appearing in the Joule–Thomson coefficient as

$$\frac{2a}{RT_i} - \frac{3abp}{R^2 T_i^2} - b = 0 \tag{9.10}$$

Calculate the inversion temperature at $p = 10.133$ MPa for N_2 gas for which $a = 0.141$ MPa dm^6 mol^{-2} and $b = 0.0392$ dm^3 mol^{-1}, $R = 0.008314$ dm^3 MPa K^{-1} mol^{-1}.
Solution: After rearranging the above equation we get,

$$(bR^2)T_i^2 - (2aR)T_i + 3abp = 0 \tag{9.11}$$

So here, the above equation looks like a quadratic equation for which N-R method can be conveniently used where T_i is a variable like x and the other acts as coefficients (Table 9.4).

Table 9.4: N-R method to solve for inversion temperature.

N	T_n	T_{n+1}	$f(x)$	$f'(x)$	ε_{abs}
1	400	−1501.246	0.3362609	0.0001768	1901.246
2	−1501.246	−566.666	−9.7945170	0.0104801	934.58
3	−566.666	−129.640	−2.3666801	0.0054154	437.026
4	−129.640	40.196	−0.5175104	0.0030471	169.836
5	40.196	76.946	−0.0781575	0.0021267	36.75
6	76.946	78.845	−0.0036595	0.0019275	1.899
7	78.845	78.850	−9.76672E-06	0.0019172	0.005

So, the temperature is 78.8 K. Similarly, giving another guess value gives 786.4 K since there are two values of T for each value of p.

9.3 Iteration method

Iteration method is also known as the fixed point iteration method. It is one of the most common and popular methods to find the real roots of a nonlinear equation or function. It is an open, simple and cyclic type of process to find roots of nonlinear equations by successive approximation. This method is used to find solution of arithmetic series, geometric series, Taylor series and many other infinite series. Also known with the name of an open bracket method or simple enclosure method, this method is somewhat slower and converges linearly like binary bisection method. But it gives quite good accuracy. It is a slow method and falls under the open method category since its convergence is not guaranteed. Its algorithm requires only one guess value and the equation is solved by the assumed approximation. It is a mathematical procedure that generates approximate solutions for a given problem in which one approximate solution is derived from the previous one. The iterations are modified at each successive level with the previous one. Iterative method provides quite accurate approximations as compared to other methods.

To find the root of nonlinear equation $f(x) = 0$ by fixed point iteration method, equation $f(x) = 0$ is written in the form of $x = \phi(x)$, that is, a new function is designed by rearranging the original function in such a manner that the highest order variables is made to be the dependent variable of x where

$$|\phi'(x)| < 1 \forall x \in (a, b) \tag{9.12}$$

where a and b are the intervals in which the root might lie.

Iteration method is valid and converges around the root in the interval which may be smaller than the interval in which $|\phi'(x)| < 1$. The convergence will take place only for a certain range of x. If the guess value is outside this range then the root will not converge. When the difference between two successive iteration is sufficiently small, iterations are stopped, that is, $|x_{n+1} - x_n| < \varepsilon_{abs}$ or until $|f(x_{n+1}) - f(x_n)| \approx 0$.

Example 4: Find a root of an equation $f(x)$ using iteration method:

$$f(x) = x^3 - x - 1 = 0 \tag{9.13}$$

Solution: Since no guess values are given here, random numbers are put in the function $f(x)$ in such way that $f(a) \times f(b) < 0$

$$f(x) = x^3 - x - 1$$

$$f(1) = (1)^3 - 1 - 1 = -1 \text{ (negative)}$$

$$f(2) = (2)^3 - 2 - 1 = 5 \text{ (positive)}$$

So the root lies between 1 and 2 (Figure 9.3). Here x^3 is rearranged to give a new function in the form $x = \phi(x)$ as

$$\phi(x) = x = (1 + x)^{1/3} \tag{9.14}$$

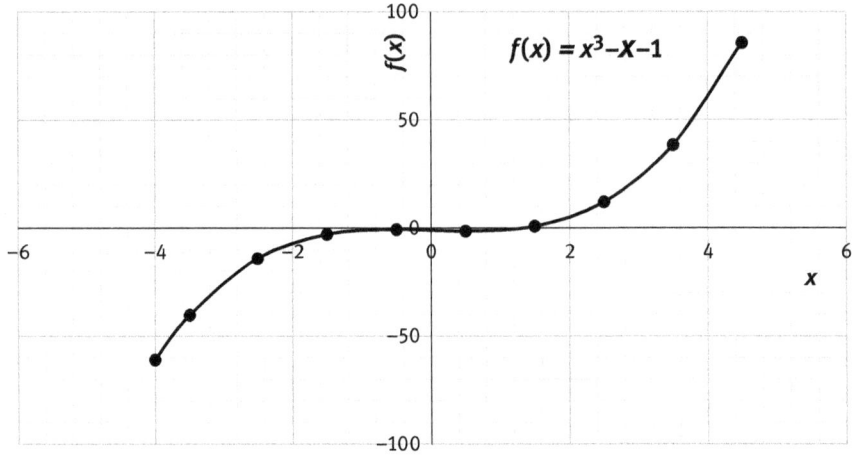

Figure 9.3: Iteration method for $f(x) = x^3-x-1$.

For iteration method only a single guess value is required which is close to the original root. Either of the guess values can be chosen from the interval in which the root lies. The results are tabulated (Table 9.5) until the function becomes zero.

Table 9.5: Iteration method for $f(x) = x^3-x-1$.

N	x_n	x_{n+1}	ε_{abs}
1	1	1.25992105	0.25992105
2	1.25992105	1.312293837	0.052372787
3	1.312293837	1.322353819	0.010059982
4	1.322353819	1.324268745	0.001914925
5	1.324268745	1.324632625	0.000363881
6	1.324632625	1.324701749	6.91233E-05
7	1.324701749	1.324714878	1.31299E-05
8	1.324714878	1.324717372	2.49399E-06

As can be seen from Table 9.5, the root is found to be 1.32471, which is accurate up to fifth decimal place in 8 iterations.

Example 5: Using eq. (9.10) used in Example 3, calculate the inversion temperature using iteration method at $p = 200$ atm for N_2 gas for which $a = 141$ kPa dm^6 mol^{-2} and $b = 0.0392$ dm^3 mol^{-1}

Solution: Rearranging the eq. (9.10),

$$T' = \sqrt{(2aRT_i - 3abP)/(bR^2)} \qquad (9.15)$$

$$1 \text{ atm} = 0.101300 \text{ MPa}$$

$$200 \text{ atm} = 20.26 \text{ MPa}$$

Using the iteration method for a guess value of 200 K (Table 9.6)

Table 9.6: Iteration method to find inversion temperature.

N	T_n	T_{n+1}	$\Delta T = T_{n+1} - T_n$
1	200	221.52	21.523
2	221.52	360.09	138.572
3	360.09	499.57	139.476
4	499.57	608.49	108.916
5	608.49	681.54	73.054
6	681.54	726.44	44.896
7	726.44	752.70	26.263
8	752.70	767.65	14.947
9	767.65	776.02	8.378
10	776.02	780.68	4.657
11	780.68	783.26	2.577
12	783.26	784.68	1.422
13	784.68	785.46	0.784
14	785.46	785.89	0.431
15	785.89	786.13	0.237
16	786.13	786.26	0.131
17	786.26	786.33	0.072
18	786.33	786.37	0.040
19	786.37	786.39	0.022

Hence in 19 iterations, the convergence has been achieved upto one decimal place and the inversion temperature is found to be 786.3K.

Example 6: Using iterative method, find the temperature at which o-toluidine has a vapour pressure of 400 mm Hg where an empirical relationship between pressure (p) and temperature (T) is given by

$$\log p = 23.8296 - \frac{3480.3}{T} - 5.081 \log T \tag{9.16}$$

Solution: At 400 mm Hg, the above equation becomes the function of temperature only.

$$21.22754 - \frac{3480.3}{T} - 5.081 \log T = 0 \tag{9.17}$$

If eq. (9.17) is rewritten as a function of T only then

$$F(T) = \frac{3480.3}{21.22754 - 5.081 \log T} = 0 \tag{9.18}$$

As evident from Table 9.7, the root converges at eighth iteration to give T = 456.9 K

Table 9.7: Iteration method to find temperature.

N	T_0	F
1	200	364.96
2	364.96	429.04
3	429.04	448.78
4	448.78	454.60
5	454.60	456.30
6	456.30	456.79
7	456.79	456.93
8	456.93	456.98

Hence at 456.9 K, o-toluidine has a vapour pressure of 400 mm Hg.

Example 7: Suppose a nuclei undergoes a radioactive decay and form a daughter nuclei which also starts undergoing radioactive decay as

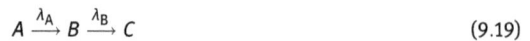

$$A \xrightarrow{\lambda_A} B \xrightarrow{\lambda_B} C \tag{9.19}$$

where λ_A and λ_B are the decay constants of nuclei A and B. The half-lives of A and B are given as 8 days and 2 days, respectively. The starting concentration of $A(N_A)$ and $B(N_B)$ are 10^{20} and zero. The variation of B with time is given by

$$N_E = \left(\frac{\lambda_A N_A^0}{\lambda_B - \lambda_E} \right) \left(e^{-\lambda_A t} - e^{-\lambda_B t} \right) \tag{9.20}$$

Find out the time at which the concentration of B is 10^{19} using Newton–Raphson method.

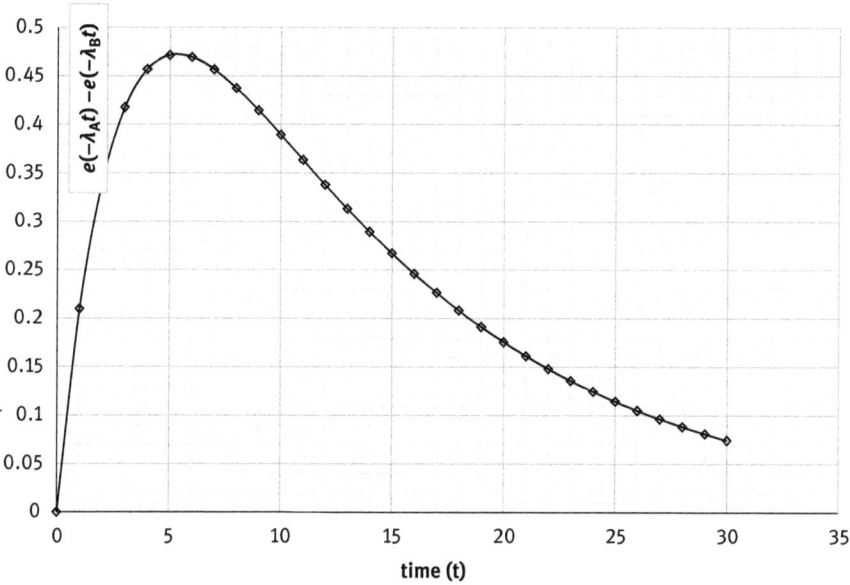

Figure 9.4: Radioactive decay.

Solution: Since the half-life and decay constants are related by the relationship is (Figure 9.4)

$$\tau = \frac{\ln 2}{\lambda} \tag{9.21}$$

which gives λ_A and λ_B as 0.0866434 and 0.346574 day^{-1}.
  Plugging all the values in the given relationship

$$N_B = \left(\frac{\lambda_A N_A^0}{\lambda_B - \lambda_A}\right)\left(e^{-\lambda_A t} - e^{-\lambda_B t}\right) \tag{9.22}$$

and rearranging gives

$$e^{-\lambda_A t} - e^{-\lambda_B t} = 0.3000 \tag{9.23}$$

which on rearranging gives a function in terms of t only as

$$f(t) = e^{-\lambda_A t} - e^{-\lambda_B t} - 0.3000 = 0 \tag{9.24}$$

The solution of the above equation can be found out using the Newton–Raphson method where the derivative is given by

$$f'(t) = -\lambda_A e^{-\lambda_A t} + \lambda_B e^{-\lambda_B t} = 0 \tag{9.25}$$

Plotting the function $e^{-\lambda_A t} - e^{-\lambda_B t}$ with time (t) shows that there are two values of time t at which the function $e^{-\lambda_A t} - e^{-\lambda_B t}$ has the value 0.3000 which are 1.6 and 13.5 days (Figure 9.4) which is also confirmed by the Newton–Raphson method (Table 9.8 and 9.9).

Table 9.8: N-R method using one guess value.

N	t	t'	f(t)	f'(t)
1	17	13.11	−0.073511115	−0.018905808
2	13.11	13.54	0.010457909	−0.024136358
3	13.54	13.54	0.000109398	-0.023624952

Table 9.9: N-R method using another guess value.

N	t	t'	f(t)	f'(t)
1	3	0.890	0.117552451	0.055721068
2	0.89	1.514	−0.108739007	0.174346804
3	1.51	1.627	−0.014659109	0.129082714
4	1.62	1.63	−0.000410034	0.121912191
5	1.63	1.63	−3.4946E-07	0.12170443

9.4 Binary bisection method

The binary bisection method (also called interval halving method) is a simple and robust root-finding algorithm that bisects an interval repeatedly in which a root is expected to lie. Hence, this method is also called bracketing. This method requires the prior knowledge of an approximate interval between which the real root lies. This method is based on the intermediate value theorem (also called Bolzano's theorem) according to which if $f(x)$ is a continuous function in the interval a and b (a and b are real numbers) such that $f(a) \times f(b) < 0$, then there is definitely a root that lies between a and b.

Let us say there is a function $f(x)$ ($f(x)$ is an algebraic or transcendental equation) which is continuous and lies in the interval a and b that has the real root, such that $f(a) \times f(b) < 0$, then at the root, that is $f(x) = 0$ (where x is the root of the equation). It is also called the zero of the function $f(x)$. In this algorithm, an interval $[a,b]$ is bisected in two (binary) and then a subinterval c is selected in which the root lies. It is halved to find a subinterval c as

$$c = \frac{a+b}{2} \tag{9.26}$$

where $c \in (a, \ b)$
    Now, find $f(c)$,
1)   If $f(c) = 0$, then c is the root of the equation, else

2) If $f(a) \times f(c) < 0$, then the root lies between a and c, so a and c becomes the new interval, hence $b = c$ and $a = a$. Again, the sub-interval c is found using the new intervals.
3) If $f(b) \times f(c) < 0$, then root lies between b and c, therefore b and c becomes the new interval, $b = b$ and $a = c$, calculate c again.

This method repeatedly bisects the new intervals based on the signs of function $f(x)$. This procedure is continued until the zero is obtained or the intervals are sufficiently small. Its convergence criteria is same as that of Newton–Raphson method and iteration method that when $f(a)$ or $f(b)$ or $f(c) \approx 0$ or $|a - b| < \epsilon_{abs}$ the iterations are stopped. The advantage of binary bisection method is that it always converges howsoever long it takes, but it may not be able to deduce more than one root at one time in a given interval. It also converges slowly as compared to other algorithms. Moreover, the interval in which we assume our root lies should be certain, else even if the interval is not true, the method converges to give some value of root which may not be true.

Example 8: Find the roots of the equation $f(x)$ using binary bisection method with $\varepsilon_{abs} = 0.01$ in the interval [1,2].

$$f(x) = x^2 - 3 \tag{9.27}$$

Solution: Here, the interval given is [1,3], that is $a = 1$, $b = 3$. Graphically, the function may be depicted as in Figure 9.5

$$f(1) = -2, f(2) = 1,$$

$$c = (1 + 2)/2 = 1.5$$

$f(c) = 1$, hence now a = a while b = c since $f(a) \times f(c) < 0$

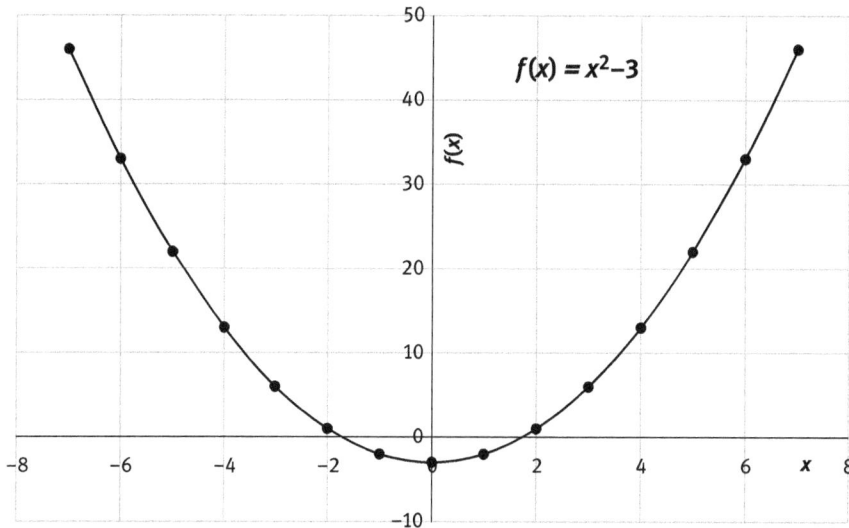

Figure 9.5: Binary bisection method for $f(x) = x^2 - 3$.

Hence, tabulating the calculations in Table 9.10 as

Table 9.10: Binary bisection method.

N	a	b	f(a)	f(b)	c = (a + b)/2	f(c)	Update
1	1	2	-2	1	1.5	-0.75	$a = c$
2	1.5	2	-0.75	1	1.75	0.062	$b = c$
3	1.5	1.75	-0.75	0.0625	1.625	-0.359	$a = c$
4	1.625	1.75	-0.3594	0.0625	1.6875	-0.1523	$a = c$
5	1.6875	1.75	-0.1523	0.0625	1.7188	-0.0457	$a = c$
6	1.7188	1.75	-0.0457	0.0625	1.7344	0.0081	$b = c$
7	1.71988	1.7344	-0.0457	0.0081	1.7266	-0.0189	$a = c$
8	1.7266	1.7344	-0.01885244	0.00814336	1.7305	-0.00536975	$a = c$
9	1.7305	1.7344	-0.00536975	0.00814336	1.73245	0.001383003	$a = c$
10	1.7305	1.73245	-0.00536975	0.001383003	1.731475	-0.001994324	$b = c$
11	1.731475	1.73245	-0.001994324	0.001383003	1.7319625	-0.000305899	$a = c$
12	1.7319625	1.73245	-0.000305899	0.001383003	1.73220625	0.000538493	$a = c$
13	1.7319625	1.73220625	-0.000305899	0.000538493	1.732084375	0.000116282	$b = c$
14	1.7319625	1.73208438	-0.000305899	0.000116299	1.73202344	-9.48033E-05	$a = c$
15	1.73202344	1.73208438	-9.48033E-05	0.000116299	1.73205391	1.07471E-05	$a = c$
16	1.73202344	1.73205391	-9.48033E-05	1.07471E-05	1.732038675	-4.20283E-05	$b = c$

Hence, in the 16th iteration, the root converges to 1.7320 which is correct up to fourth decimal place.

9.5 Secant method

As the name suggests, secant implies a line which passes through two points of the curve. It is an algorithm for finding the roots of scalar-valued function of a single variable x when no information of derivative is given. Secant method is an algorithm used to find the roots of nonlinear functions. Let x_0 and x_1 be the two initial guesses for the root of $f(x) = 0$, then $f(x_0)$ and $f(x_1)$, respectively, are their function values. A line is drawn between the two guess approximations and the point where it crosses the x-axis (Figure 9.6) will be the approximate root of $f(x)$ using approximate guesses (x_0 and x_1). Secant method assumes that the equation or function is approximately linear in the region of interest.

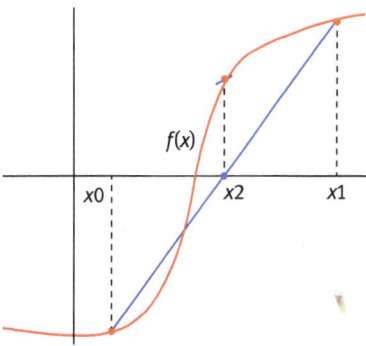

Figure 9.6: Secant method.

Hence, the slope m can be written as

$$m = \frac{y - f(x_1)}{x - x_1} = \frac{f(x_1) - f(x_0)}{x_1 - x_0} \tag{9.28}$$

So, in the above figure, it can be seen that the line between the two guess values touches x-axis at x_2, so here, value of $y = 0$ (root is the value of a variable where the function becomes zero); therefore

$$m = \frac{0 - f(x_1)}{x_2 - x_1} = \frac{f(x_1) - f(x_0)}{x_1 - x_0} \tag{9.29}$$

$$-f(x_1) = \frac{f(x_1) - f(x_0)}{x_1 - x_0}(x_2 - x_1) \tag{9.30}$$

which on rearranging gives

$$x_2 = x_1 - f(x_1) \frac{x_1 - x_0}{f(x_1) - f(x_0)} \quad (9.31)$$

Hence, x_2 is the approximate root yet not the original root. A root for the function $f(x)$ is approximated which is not really a root but is quite near to the original root.

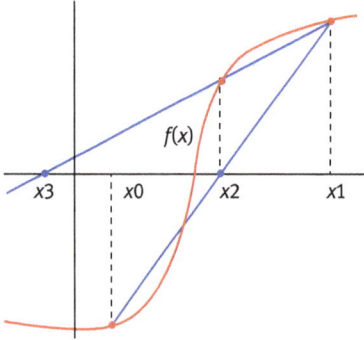

Figure 9.7: Secant method.

To do the same, a new secant would be drawn (Figure 9.7). Since x_2 is closer to the original root than x_0 (also $f(x_1)^*f(x_2) < 0$), the new secant would be drawn between new intervals, that is $[x_1, f(x_1)]$ and $[x_2, f(x_2)]$ where x_3 is the new point of intersection to x-axis. Using the same procedure as above, it can be written that

$$x_3 = x_2 - f(x_2) \frac{x_2 - x_1}{f(x_2) - f(x_1)} \quad (9.32)$$

This cycle will continue to take place until the difference between $|x_{n-1} - x_{n-2}| < \varepsilon_{abs}$. Hence, the general expression for secant method may be written as

$$x_n = x_{n-1} - f(x_{n-1}) \frac{x_{n-1} - x_{n-2}}{f(x_{n-1}) - f(x_{n-2})} \quad (9.33)$$

In above procedure, the iterations are used to bring the two guess values close to the original root. The most important point to remember is that the initial guess value should be close to the original root else secant method will not converge. There may not be a proper way of knowing how close the guess value may be but if the function $f(x)$ is differentiable and $f'(x_0) = 0$ on that interval then the secant method would not converge. Thus, it adopts the possibility of being faster but may not be converging due to non-root bracketing.

As compared to N-R method and iteration method, binary bisection method and secant method uses two guess values (interval in which root lies). As compared to N-R method, where $f(x)$ and $f'(x)$ both are calculated, secant method should be faster since it has to compute only one $f(x)$, but in reality N-R method is found to be faster as compared to secant method since $f(x)$ and $f'(x)$ are computed simultaneously as it iterates at each step. Secant method may not necessarily always converge.

Example 9: Using secant method, find roots of the following equation upto three decimal places

$$f(x) = x^3 - x - 1 \tag{9.13}$$

Solution: Using Example 4, the roots lies between 1 and 2, hence using secant formula (9.32), the first iteration be $N = 1$

$$x_n = x_{n-1} - f(x_{n-1}) \frac{x_{n-1} - x_{n-2}}{f(x_{n-1}) - f(x_{n-2})} \tag{9.33}$$

$$x_1 = 2 - 5\left(\frac{2-1}{5-(-1)}\right)$$

$$x_1 = 1.166$$

The remaining results are tabulated (Table 9.11) as follows.

Table 9.11: Secant method for solving f(x) = x³−x − 1.

N	x_{n-2}	x_{n-1}	$f(x_{n-2})$	$f(x_{n-1})$	x_n	$f(x_n)$
1	1	2	−1	5	1.166666	−0.578703
2	1.166666	2	−0.578703	5	1.253112	−0.285363
3	1.253112	2	−0.285363	5	1.293437	−0.129542
4	1.293437	2	−0.129542	5	1.311281	−0.056588
5	1.311281	2	−0.056588	5	1.318988	−0.024303
6	1.318988	2	−0.024303	5	1.322282	−0.010361
7	1.322282	2	−0.010361	5	1.323684	−0.004403
8	1.323684	2	−0.004403	5	1.324279	−0.001869
9	1.324279	2	−0.001869	5	1.324531	−0.000792

The root is found to be 1.324 which is exact upto three decimal place which is same as obtained using iteration method.

Example 10: Find the root of the nonlinear equation $3x + \sin x - e^x = 0$

Solution: When analysed graphically, both the root lies between [0, 1]. So, using the secant formula as (Figure 9.8)

$$x_n = x_{n-1} - f(x_{n-1}) \frac{x_{n-1} - x_{n-2}}{f(x_{n-1}) - f(x_{n-2})} \tag{9.33}$$

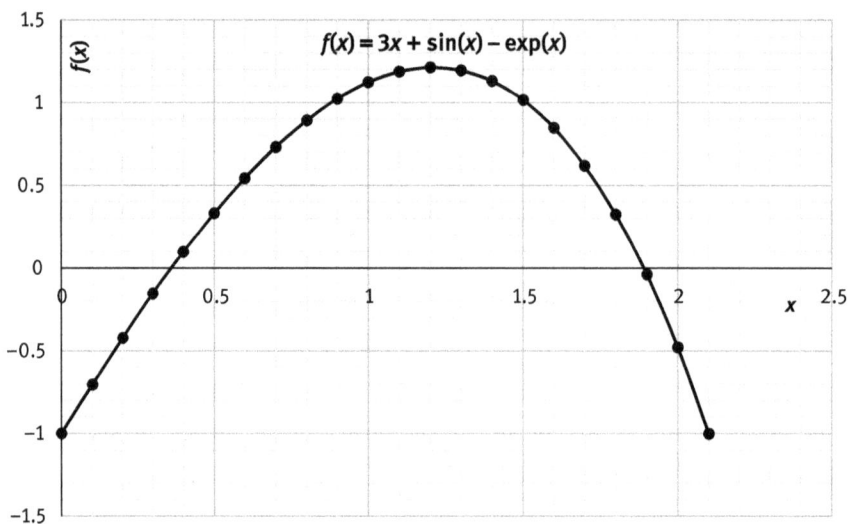

$$f(x) = 3x + \sin(x) - \exp(x)$$

Figure 9.8: Secant method.

The results are tabulated in Table 9.12:

Table 9.12: Secant method for $f(x) = 3x + \sin(x) - e^x$.

N	x_{n-2}	x_{n-1}	$f(x_{n-2})$	$f(x_{n-1})$	x_n
1	0	1	−1	1.123189156	0.470989595
2	1	0.470989595	1.123189156	0.265158816	0.307508461
3	0.470989595	0.307508461	0.265158816	−0.134822017	0.362613242
4	0.307508461	0.362613242	−0.134822017	0.005478529	0.360461482
5	0.362613242	0.360461482	0.005478529	9.95177E-05	0.360421672
6	0.360461482	0.360421672	9.95177E-05	−7.80142E-08	0.360421703

So here, after six iterations, the root is found to be 0.3604, which is exact up to four decimal places. The iterations can be continued until a convergence criteria is reached which is $|x_{n+1} - x_n| < \varepsilon_{step}$, that is, the difference between successive iterates is sufficiently small. If the denominator is zero, then division by zero fails the iteration. Hence there is a need for new initial approximation. If after a given number of iterations, convergence is not reached, then it is possible that solution (root) does not exist.

9.6 Regula-Falsi method

Regula-Falsi method is also a numerical method to find roots of equation. As the name suggests it means a false position method, which approximates a real root by approximating a guess root in an interval which lies close to the original root. It is also a bracketing method like binary bisection method to find roots of the equation and hence it is always bound to converge. If the function $f(x)$ is continuous and there is an interval in which the root lies of the function, then the root of the function can be found using R-F method. This method is very similar to binary bisection method except in finding the new intervals. Also, it differs from binary bisection method in the speed of convergence, R-F converges faster than binary bisection method.

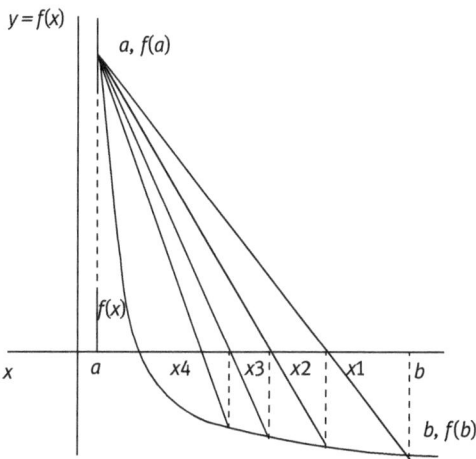

Figure 9.9: Regula-Falsi method.

This method alike other methods also assumes that the function $f(x)$ be continuous. Regula-Falsi method begins with a and b interval (such that $a < z < b$ where z is the root of the $f(x)$) such that $f(a) \times f(b) < 0$, then to find the root a new approximate root x_1 is calculated which is not a real root but it is a false root as suggested by the name and proceed to find an approximate root in the interval chosen and then carefully selecting which roots should be taken forward for the next iteration (either a and x_1 or x_1 and b) depending upon the sign of $f(x_1)$, $f(a)$ and $f(b)$. This is called root bracketing, implying finding root within the interval chosen. For this, a line is drawn between $(a, f(a))$ and $(b, f(b))$ which is known as interpolation line. The new false root is checked for its magnitude of the function, that is $f(x_1)$.

There may be three possibilities:
1) If $f(x_1) = 0$, then x_1 is the root but this usually does not happen.
2) If $f(a) \times f(x_1) < 0$, then x_1 becomes the new b, that is, $b = x_1$, so now an interpolation line is drawn between $a, f(a)$ and $x_1, f(x_1)$.
3) If $f(b) \times f(x_1) < 0$ then x_1 becomes the new a, that is, $a = x_1$ and a new interpolation line is drawn between $x_1, f(x_1)$ and $b, f(b)$ (Figure 9.9).

When a chord is drawn between $[a, f(a)]$ and $[b, f(b)]$, the general equation for the line is

$$\frac{y - y_1}{x - x_1} = \frac{y_2 - y_1}{x_2 - x_1} \tag{9.34}$$

$$\frac{y - f(a)}{x - a} = \frac{f(b) - f(a)}{b - a} \tag{9.35}$$

At $x = x_1, y = 0$

$$\frac{0 - f(a)}{x_1 - a} = \frac{f(b) - f(a)}{b - a} \tag{9.36}$$

$$x_1 = \frac{af(b) - bf(a)}{f(b) - f(a)} \tag{9.37}$$

Here x_1 is the first approximated root, since it is still not the original root. Check if $f(a) \times f(x_1) < 0$, then $b = x_1$ else $a = x_1$ ($f(b) \times f(x_1)$). Here $f(a) \times f(x_1) < 0$ so, a new chord is drawn between $(a, f(a))$ and $(x_1, f(x_1))$ and the same equation of line is written as

$$\frac{y - f(a)}{x - a} = \frac{f(x_1) - f(a)}{x_1 - a} \tag{9.38}$$

At $x = x_2, y = 0$

$$\frac{0 - f(a)}{x_2 - a} = \frac{f(x_1) - f(a)}{x_1 - a} \tag{9.39}$$

$$x_2 = \frac{af(x_1) - x_1 f(a)}{f(x_1) - f(a)} \tag{9.40}$$

This above procedure is repeated $(x_2, x_3, x_4, \ldots, x_n)$ until $f(x_n) = 0$ or $\approx \varepsilon_{abs}$ (also called tolerance limit or degree of accuracy).

Regula-Falsi methods resemble the secant method in the formula used and two starting initial approximations, except secant method. Except, secant method may not necessarily find the approximated root in the starting interval (x_0 and x_1). Hence roots are not bracketed in the secant method so it may not always converge but Regula-Falsi method always converge.

Example 11: Find the roots of the function $f(x)$ using Regula-Falsi method in the interval [0,2]. Given $\varepsilon_{abs} = 0.0001$

$$f(x) = x^3 + 3x - 5 \tag{9.41}$$

Solution: Graphically, the function $f(x)$ is depicted in Figure 9.10. For approximating the first order, we will use eq. (9.37)

$$x_1 = \frac{af(b) - bf(a)}{f(b) - f(a)} \tag{9.37}$$

$$f(a) = f(0) = -5, \quad f(b) = f(2) = 9$$

$$x_1 = \frac{0f(2) - 2f(0)}{f(2) - f(0)} \tag{9.38}$$

$$x_1 = \frac{5}{7} = 0.7142$$

Now $f(x_1) = -2.492$, since $f(b) \times f(x_1) < 0$, hence $a = x_1$, the further results are tabulated in Table 9.13.

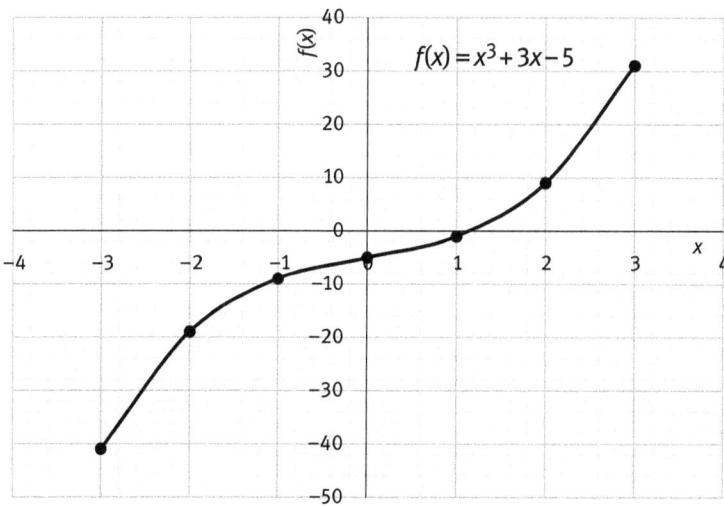

Figure 9.10: Regula-Falsi method for $f(x) = x^3 + 3x - 5$.

Table 9.13: Regula-Falsi method for $f(x) = x^3 + 3x - 5$.

N	a	b	f(a)	f(b)	x_n	$f(x_n)$	Update
1	0	2	-5	9	0.714285714	-2.49271137	$a = x$
2	0.714285714	2	-2.49271137	9	0.993150685	-1.040955472	$a = x$
3	0.993150685	2	-1.040955472	9	1.097531718	-0.38534462	$a = x$
4	1.097531718	2	-0.38534462	9	1.13458537	-0.135710338	$a = x$
5	1.13458537	2	-0.135710338	9	1.147441044	-0.046931951	$a = x$
6	1.147441044	2	-0.046931951	9	1.151863787	-0.016127073	$a = x$
7	1.151863787	2	-0.016127073	9	1.153380841	-0.005529515	$a = x$
8	1.153380841	2	-0.005529515	9	1.153900677	-0.001894482	$a = x$
9	1.153900677	2	-0.001894482	9	1.154078742	-0.000648906	$a = x$
10	1.154078742	2	-0.000648906	9	1.154139729	-0.000222246	$a = x$
11	1.154139729	2	-0.000222246	9	1.154160616	-7.61155E-05	$a = x$
12	1.154160616	2	-7.61155E-05	9	1.154167769	-2.6068E-05	$a = x$

From the tabulated results it is clear that at the 12th iteration, the root (1.15416) is stable up to 5th decimal place. Also at 12th iteration $f(x_1) \approx 10^{-5}$

Example 12: Determine the volume of 0.5 mol of the van der Waal's gas CO_2 at 2226 kPa and 298 K. Given $a = 363.76$ kPa dm^6 mol^{-2} and $b = 42.67$ cm^3 mol^{-1}.

Solution: The guess value for volume is taken using ideal gas equation for volume as $V = nRT/p$

$$\left(p + \frac{n^2 a}{V^2}\right)(V - nb) = nRT \tag{9.39}$$

which on rearranging gives

$$f(V) = V^3 - \left(nb + \frac{nRT}{p}\right)V^2 + \frac{n^2 a}{p}V - \frac{n^3 ab}{p} = 0 \tag{9.40}$$

To apply N-R method, derivative of function is required as

$$f'(V) = 3V^2 - 2\left(nb + \frac{nRT}{p}\right)V + \frac{n^2 a}{p} \tag{9.41}$$

Hence, the results are tabulated in Table 9.14.

Table 9.14: N-R method to solve van der Waal's equation.

N	x_0	f(x)	f'(x)	x_n
1	0.55	0.013175444	0.312726648	0.507869136
2	0.507869136	0.001828307	0.227709846	0.499840028
3	0.499840028	6.04526E-05	0.21271595	0.499555834

0.4998 dm^3 is the volume of CO_2 obtained for the given condition. It is important to note that the order of polynomial is always equal to the number of roots of that polynomial. Here, since the order of V is 3, it implies that there should be three roots but it is not possible that at a single pressure, that is, 2,226 kPa, there would be three volumes, hence only one of them is true. In this case, one may obtain two more volumes (using different guess values) which would not be true solution to the equation.

Example 13: The concentration of $[H^+]$ in a dilute solution of HCl having concentration 10^{-6} M is given by the expression

$$[H^+]^2 - [HCl][H^+] - k_w = 0 \tag{9.42}$$

Calculate the pH of 10^{-7} M HCl solution at 25 °C

Solution: The equation given above is a quadratic equation of the type $ax^2 + bx + c = 0$. The equation can be solved using iteration method by rearranging the equation as

$$[H^+] = \sqrt{[HCl][H^+] + k_w} \tag{9.43}$$

Taking the initial guess of the acid as 10^{-7}, the results are tabulated in Table 9.15.

Table 9.15: Iteration method to solve for pH.

N	x_n	x_{n+1}	E_{abs}
1	0.0000001	3.16228E-07	2.16228E-07
2	3.16228E-07	5.62341E-07	2.46114E-07
3	5.62341E-07	7.49894E-07	1.87553E-07
4	7.49894E-07	8.65964E-07	1.1607E-07
5	8.65964E-07	9.30572E-07	6.46077E-08
6	9.30572E-07	9.64662E-07	3.40896E-08
7	9.64662E-07	9.82172E-07	1.75103E-08
8	9.82172E-07	9.91046E-07	8.87397E-09
9	9.91046E-07	9.95513E-07	4.467E-09

$$pH = -\log_{10}[H^+]$$ (9.44)

$$pH = -\log_{10}[9.9 \times 10^{-7}]$$ (9.45)

$$pH = 6.00$$

9.7 Problems for practice

1. Using the Newton–Raphson method, find the root of the equation accurate to within 10^{-7}
   (a) $5\sin x^2 - 8\cos^5 x = 0$ for $[-1,2]$ (b) $x^4 - x - 10 = 0$ for $[1,2]$

2. Find the root of the equation $x^2 - 4x - 7 = 0$ near $x = 5$ to the nearest thousandth.

3. Find the roots of the following functions using iteration method
   (a) $f(x) = x^2 - 2$ starting with $x_0 = 1$
   (b) $f(x) = e^{-x}\cos(x)$ starting with $x_0 = 1.3$ where $\varepsilon_{abs} = 10^{-5}$

4. Using the iteration method, calculate $[H^+]$ in 10^{-4} M solution of weak acid with $k_a = 1.7 \times 10^{-10}$ M at 25 °C. Given the equation as

$$[H^+]^3 + [H^+]^2 k_a - [H^+]\{k_w + k_a[HA]_0\} - k_a k_w = 0$$

   Find the root of the equation $x^3 - 3$ with the bisection method with the interval [1, 2] using $\varepsilon_{step} = 0.1$ and $\varepsilon_{abs} = 0.1$

5. Using binary bisection method, find the roots of the following function.
   (a) $\cos(x) - xe^x = 0$ (b) $x^4 - x - 10 = 0$ (c) $x - e^{-x} = 0$

6. Find root of the $f(x) = x \times \cos[(x)/(x-2)] = 0$ by Regula-Falsi method.

7. Find the root of the function $f(x) = 5\sin^2 x - 8\cos^5 x = 0$ in the interval [0.5,1.5]

8. Find root of the equation $x^4 - x - 10 = 0$ using secant method in the interval [1,2]

9. Find root of the following equations using secant method
   (a) $\exp(-x) = 3\log(x)$ (b) $x\sin(x) - (1/2) = 0$

Chapter 10
Differentials and integrals

10.1 Introduction

There are two types of variables: dependent variables and independent variables. While the dependent variables are dependent on the estimation of another variable in its condition, the independent variables are independent of any changes, that is, values of independent variables are not dependent on any changes. For example, in the function $y = 2x + 7$, x is an independent variable and the value of y is a function of x value, hence dependent on x. It can also be represented as

$$y = f(x) = 2x + 7 \qquad (10.1)$$

In calculus also, there are two categories namely differential calculus (derivative) and integral calculus. While the derivative measures the change in dependent variable upon changing independent variables (also called the rate of change), the integral measures the area under the curve. In this chapter, both have been discussed and also their utility in chemical thermodynamics.

10.2 Partial derivatives

An ordinary derivative like $df(x)/dx$ can be interpreted as a change in function $f(x)$ as $df(x)$ when there is a slight change in x, that is, dx. It can also be understood as the slope of the graph. There is no other way than to understand that only x is able to vary the function $f(x)$. But sometimes, there are more than two variables that affect the function which can be written as $f(x,y)$.

For instance, consider

$$f(x,y) = x^2 + 5xy \qquad (10.2)$$

Here $df(x,y)/dx = 2x + 5y$ will only represent the change in function due to the change in x but the change in function due to change in y is not considered.

Similarly, $df(x,y)/dy$ only tells the change in function on slight change of dy. Not a single derivative would be able to explain the overall change in $f(x,y)$ since our function $f(x,y)$ is a two-dimensional dependent function and may change the input in two directions. That is why these are called partial derivatives. So instead of using the d symbol, another new symbol ∂ is used to represent the partial derivative. There are as many derivatives as the number of variables keeping the others constant while differentiating one at a time.

https://doi.org/10.1515/9783111334448-010

For example,

$$df(x,y) = \left(\frac{\partial f(x,y)}{\partial x}\right)dx + \left(\frac{\partial f(x,y)}{\partial y}\right)dy \tag{10.3}$$

Likewise in thermodynamics (ideal gas), P is a function of T, V and n as $P(T, V, n)$:

$$dP = \left(\frac{\partial P}{\partial T}\right)dT + \left(\frac{\partial P}{\partial V}\right)dV + \left(\frac{\partial P}{\partial n}\right)dn \tag{10.4}$$

Here $(\partial P/\partial T)$, $(\partial P/\partial V)$ and $(\partial P/\partial n)$ are called partial derivatives w.r.t T, V and n, respectively.

10.2.1 Reciprocal identity

According to the reciprocal identity, a derivative (dy/dx) is equal to the reciprocal of the derivative, where the dependent (y) and independent (x) variables are reversed keeping z constant in both the derivatives as follows:

$$\left(\frac{\partial y}{\partial x}\right)_z = \left(\frac{1}{(\partial x/\partial y)_z}\right) \tag{10.5}$$

Example 1: Show that for an ideal gas,

$$\left(\frac{dP}{dV}\right)_{n,T} = \left(\frac{1}{(dV/dP)_{n,T}}\right) \tag{10.6}$$

Solution: For an ideal gas $PV = nRT$.
From the above, differentiate P w.r.t V and then V w.r.t P as

$$P = nRT/V \tag{10.7}$$

$$\left(\frac{dP}{dV}\right)_{n,T} = \frac{-nRT}{V^2} \tag{10.8}$$

$$\left(\frac{dP}{dV}\right)_{n,T} = \frac{-PV}{V^2} = \frac{-P}{V} \tag{10.9}$$

and

$$V = nRT/P \tag{10.10}$$

$$\left(\frac{dV}{dP}\right)_{n,T} = \frac{-nRT}{P^2} \tag{10.11}$$

$$\left(\frac{dV}{dP}\right)_{n,T} = \frac{-PV}{P^2} = \frac{-V}{P} \tag{10.12}$$

Hence,

$$\left(\frac{dP}{dV}\right)_{n,T} = \left(\frac{1}{(dV/dP)_{n,T}}\right) \tag{10.13}$$

10.2.2 Second partial derivative

Consider z as a function of x variable, then it can be differentiated twice *w.r.t* x as

$$\left(\frac{\partial^2 z}{\partial x^2}\right) = \left[\frac{\partial}{\partial x}\left(\frac{\partial z}{\partial x}\right)\right] \tag{10.14}$$

But if z is a function of x and y (as $z = f(x,y)$) then the second partial derivative keeping y constant can be represented as

$$\left(\frac{\partial^2 z}{\partial x^2}\right) = \left[\frac{\partial}{\partial x}\left(\frac{\partial z}{\partial x}\right)_y\right]_y \tag{10.15}$$

10.2.3 Mixed second partial derivative

When z is the function of two variables, then z is differentiated *w.r.t* x keeping y constant, and then further differentiating *w.r.t* y keeping x constant as

$$\left(\frac{\partial^2 z}{\partial y \partial x}\right) = \left[\frac{\partial}{\partial y}\left(\frac{\partial z}{\partial x}\right)_y\right]_x \tag{10.16}$$

These are classified as mixed second partial derivative.

10.2.4 Euler's reciprocity relationship

According to Euler's reciprocity relationship, if z is a function of both x and y, then their mixed second partial derivatives must be equal to each other as

$$\left(\frac{\partial^2 z}{\partial x \partial y}\right) = \left(\frac{\partial^2 z}{\partial y \partial x}\right) \tag{10.17}$$

whether z is differentiated w.r.t x first, then y and vice versa.

10.2.5 Cyclic rule

The cyclic rule, also called the triple product, relates partial derivatives of three interdependent variables as

$$\left(\frac{\partial z}{\partial x}\right)\left(\frac{\partial y}{\partial z}\right)\left(\frac{\partial x}{\partial y}\right) = -1 \tag{10.18}$$

The cyclic rule allows to find the partial derivatives that are otherwise difficult to evaluate or measure by rearranging the derivatives.

For example,

$$\left(\frac{\partial z}{\partial x}\right) = \frac{\left(\frac{\partial y}{\partial x}\right)}{\left(\frac{\partial y}{\partial z}\right)} \tag{10.19}$$

This cyclic rule is useful in thermodynamics, where one variable is usually a function of another two variables. For instance, the thermodynamic equation of state $PV = nRT$ relates three variables P, V and T:

$$P = f(V, T) \tag{10.20}$$

Then according to the cyclic rule

$$\left(\frac{\partial P}{\partial T}\right)_V \left(\frac{\partial T}{\partial V}\right)_P \left(\frac{\partial V}{\partial P}\right)_T = -1 \tag{10.21}$$

Example 2: Prove the cyclic rule for $PV = nRT$.

Solution: Evaluating each partial derivative as

$$\left(\frac{\partial P}{\partial T}\right)_V = \frac{nR}{V} \tag{10.22}$$

$$\left(\frac{\partial T}{\partial V}\right)_P = \frac{P}{nR} \tag{10.23}$$

$$\left(\frac{\partial V}{\partial P}\right)_T = \frac{-nRT}{P^2} \tag{10.24}$$

$$\left(\frac{\partial P}{\partial T}\right)_V \left(\frac{\partial T}{\partial V}\right)_P \left(\frac{\partial V}{\partial P}\right)_T = \left(\frac{nR}{V}\right)\left(\frac{P}{nR}\right)\left(\frac{-nRT}{P^2}\right) = -1 \tag{10.25}$$

which proves the cyclic rule.

10.3 Stationary points

A stationary point is a point where the derivative of the function becomes zero. At the stationary point, the function could be maximum, minimum or a point of inflection. It can be interpreted that the slope of the function is zero at the stationary point, hence the name stationary. The maximum or minimum points (where the derivative changes the sign from being positive to negative or vice versa) are also called a *turning point*.

A stationary point can be a local maximum or a local minimum depending upon its second derivative. For a single variable function, a stationary point is a local maximum (where the function is maximum) if its second derivative $f''(x) < 0$, whereas it is a local minimum (where the function has the smallest value) if $f''(x) > 0$. Hence, it can be said that all the turning points are stationary points but not vice versa. For example, if the stationary point lies at $x = a$, then if $f''(x) < 0$ at $x = a$, it is a maximum, whereas if $f''(x) > 0$ at $x = a$, it is a minimum (Tab 10.1).

Table 10.1: Nature of stationary points based on their derivative $f'(x)$.

	Maximum	**Minimum**	**Point of inflection**
Before	$f'(x)$ = positive	$f'(x)$ = negative	$f'(x)$ = positive
After	$f'(x)$ = negative	$f'(x)$ = positive	$f'(x)$ = positive

The points of inflection are the stationary points where the graph of the function changes from being concave to convex (rising point) and vice versa (falling point). It can be said that at the point of inflection, the slope of the function is neither increasing nor decreasing (Fig 10.1). It can be evaluated by finding the second derivative of the function which if turns out to be zero, then it is a point of inflection. If $f''(x) = 0$, it is called the point of inflection. To know if the point of inflection is either rising or falling, the third derivative sign is used. If $f'''(x) > 0$ then it is a rising point of inflection, and if $f'''(x) < 0$, it is the falling point of inflection.

Figure 10.1: Nature of stationary points.

Saddle point is a point of inflection only when multivariable are involved. It is a point on a surface (a two-dimensional surface in three dimensions) where the tangent plane is horizontal and the function is neither maxima nor minima. Similarly, for the function with two variables, the condition for stationary point is maximum if f_{xx} (or f_{yy}) < 0 and minimum if f_{xx} (or f_{yy}) > 0. Also if $f_{xx} f_{yy} - f_{xy}^2 > 0$ for either a maximum or minimum and if $f_{xx} f_{yy} - f_{xy}^2 < 0$, then it is a saddle point. A saddle point is a stationary point but does not need to be an extremum. It can be checked by extremum test of $f'(x) = 0$ while $f''(x) > 0$.

For example, in one dimension,

$$f(x) = x^3 \tag{10.26}$$

$$f'(x) = 3x^2 \tag{10.27}$$

$$f''(x) = 6x \tag{10.28}$$

$$f'''(x) = 6 \tag{10.29}$$

Here at $x_0 = 0$, the value of $f''(x_0) = 0$ while $f'''(x_0) = 6$.

Example 3: Find the stationary point and determine its nature:

$$f(x) = x^4 + 4x^3 - 2 \tag{10.30}$$

Solution: From above, we learnt that at the stationary point, $f'(x) = 0$ implying

$$f'(x) = 4x^3 + 12x^2 = 0 \tag{10.31}$$

which leaves us with the solution $x_1 = -3$ and $x_2 = 0$. Hence, these are the stationary points. To check the nature of the stationary point, one needs to find the second derivative as follows:

$$f''(x) = 12x^2 + 24x \tag{10.32}$$

$f''(-3) = 36 > 0$, hence x_1 is the local mimima.

At $x_2 = 0$, $f''(x) = 0$. Hence, its nature cannot be justified by the second derivative test. Therefore, its first derivative test must be used:

$$f'(x) = 4x^3 + 12x^2 = 0 \tag{10.31}$$

According to the above equation of first derivative,

$$\text{if } x < 0 \text{ then } f'(x) < 0$$

and

$$\text{if } x > 0, \text{ then } f'(x) > 0$$

which implies that on either side of $x = 0$, the function is increasing and decreasing. Hence $x_2 = 0$ is the point of inflection.

In gaseous state, pV isotherm of a gas (Fig 10.2) shows a stationary inflection point at X, where both $(\partial p/\partial V)_T = 0$ and $(\partial^2 p/\partial V^2)_T = 0$. X is a critical point where only one phase exists and the heat of vaporization is zero.

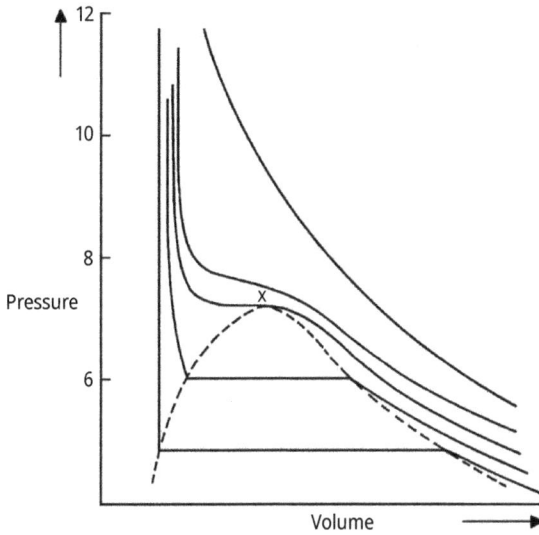

Figure 10.2: pV isotherm showing the point of inflection at critical point (X).

10.4 Total differentials

The infinitesimal change dy is given by the expression called total differential of y. Whenever there are two variables for a given function, its total differential is equal to the sum of each partial derivative while keeping the other variables constant. For instance, y is a function of different x variables as

$$y = f(x_1, x_2, x_3, \ldots, x_n) \tag{10.33}$$

then

$$dy = \left(\frac{\partial A}{\partial x_1}\right)dx_1 + \left(\frac{\partial A}{\partial x_2}\right)dx_2 + \left(\frac{\partial A}{\partial x_3}\right)dx_3 + \cdots + \left(\frac{\partial A}{\partial x_n}\right)dx_n \tag{10.34}$$

Example 4: Find the total differential of the given function

$$z = xy + \cos y \tag{10.35}$$

Solution:

$$dz = \left(\frac{\partial z}{\partial x}\right)_y dx + \left(\frac{\partial z}{\partial y}\right)_x dy \tag{10.36}$$

$$dz = (y)dx + (x - \text{Sin } y)dy \tag{10.37}$$

Example 5: If $P = f(V,T)$, then write its total derivative.

Solution: If an infinitesimal change is made in volume as dV and in temperature as dT keeping n fixed, then the change in pressure can be written as

$$dP = \left(\frac{\partial P}{\partial T}\right)_{V,n} dT + \left(\frac{\partial P}{\partial V}\right)_{T,n} dV \tag{10.38}$$

It can be seen that each term of this equation is the change in one independent variable and each partial derivative is taken with the other independent variables treated as constants. Here, dP is called the total differential of P.

10.4.1 Exact and inexact integrals

Consider the function z as

$$z = a^3 + a^2b + ab^2 \tag{10.39}$$

then its total differential can be written as

$$dz = \left(3a^2 + 2ab + b^2\right)da + \left(a^2 + 2ab\right)db \tag{10.40}$$

which can also be written in general differential form as

$$dz = X(a,b)da + Y(a,b)db \tag{10.41}$$

If this is the differential of a function (exact differential), then X and Y must be the appropriate derivatives of that function. But if they are not the appropriate differentials of the function, then in that case dz is called an inexact differential. It can be calculated from specific values of da and db but is not equal to the change in any function of a and b resulting from these changes.

In order to differentiate between exact and inexact differentials, Euler's reciprocity relation is used. If the function z exists as $z(a,b)$, such that

$$X(a,b) = \left(\frac{\partial z}{\partial a}\right)_b \text{ and } Y(a,b) = \left(\frac{\partial z}{\partial b}\right)_a \tag{10.42 and 10.43}$$

then according to Euler's reciprocity relationship,

$$\left(\frac{\partial^2 z}{\partial b \partial a}\right) = \left(\frac{\partial^2 z}{\partial a \partial b}\right) \tag{10.44}$$

implying that the differential is exact and the order of differentiation is immaterial. Also it can be written as

$$\left(\frac{\partial X}{\partial b}\right)_a = \left(\frac{\partial Y}{\partial a}\right)_b \qquad (10.45)$$

Hence, for a differential to be exact, Euler's reciprocity relationship should be obeyed.

Example 6: Prove that the following differential is exact:

$$z = 3x^2y^3\,dx + 3x^3y^2\,dy \qquad (10.46)$$

Solution: If the differential is exact, then it should obey Euler's reciprocity relation as

$$A(x,y)dx + B(x,y)dy = 3x^2y^3\,dx + 3x^3y^2\,dy \qquad (10.47)$$

Hence,

$$\left(\frac{\partial A}{\partial y}\right) = \left(\frac{\partial\left(3x^2y^3\right)}{\partial y}\right) = 9x^2y^2 \qquad (10.48)$$

and

$$\left(\frac{\partial B}{\partial x}\right) = \left(\frac{\partial\left(3x^3y^2\right)}{\partial x}\right) = 9x^2y^2 \qquad (10.49)$$

$$\left(\frac{\partial A}{\partial y}\right)_a = \left(\frac{\partial B}{\partial x}\right)_y \qquad (10.50)$$

Heat as inexact differential

In chemical thermodynamics, heat is a path function and hence can be proved as an inexact differential. According to the first law, if we restrict ourselves to the work done as mechanical pV work, then

$$dq = dU - dw \qquad (10.51)$$

$$dq = dU + pdV \qquad (10.52)$$

Since internal energy U is a state function and is a function of V and T, $U = f(V,T)$, then its total differential is

$$dU = \left(\frac{\partial U}{\partial V}\right)_T dV + \left(\frac{\partial U}{\partial T}\right)_V dT \qquad (10.53)$$

On substituting the above equation in heat equation, we have

$$dq = \left(\frac{\partial U}{\partial V}\right)_T dV + \left(\frac{\partial U}{\partial T}\right)_V dT + pdV \qquad (10.54)$$

Dividing the above equation by dV and applying the constant T condition, we have

$$dq = \left(\frac{\partial U}{\partial V}\right)_T dV + pdV \qquad (10.55)$$

Similarly dividing the above equation by dT and applying constant V condition, we have

$$dq = \left(\frac{\partial U}{\partial T}\right)_V dT \qquad (10.56)$$

Only if q was an exact differential, it must follow the Euler's reciprocity relationship as

$$\left(\frac{\partial^2 q}{\partial V \partial T}\right) = \left(\frac{\partial^2 q}{\partial T \partial V}\right)$$

(10.57)

Applying the same relation above and evaluating the second differentials, we get

$$\left(\frac{\partial^2 U}{\partial T \partial V}\right) + \left(\frac{\partial p}{\partial T}\right)_V = \left(\frac{\partial^2 U}{\partial V \partial T}\right)$$

(10.58)

Since U is already an exact differential

$$\left(\frac{\partial^2 U}{\partial T \partial V}\right) = \left(\frac{\partial^2 U}{\partial V \partial T}\right)$$

(10.59)

which leaves

$$\left(\frac{\partial p}{\partial T}\right)_V = 0$$

(10.60)

The above equation means that there is no change in pressure when temperature changes at constant volume which is contrary to Charles's law. Hence, the assumption of q as an exact differential is wrong and it is an inexact differential.

Similarly, it can also be proved that the work done is also an inexact differential and hence the path function.

10.5 Line integrals

A line integral is an integral in which the function is to be integrated along a curve also called as path integral or curvilinear integral (Fig 10.3).

Consider a function $F(x)$ and the integral

$$\int_a^b F(x)\,dx$$

(10.61)

In the earlier chapter, we learnt that integral from a single variable (x) can be understood as an area under the curve or sum of infinitesimal increments where function $F(x)$ is moving along x axis from a to b.

Alternatively if the function $F(x)$ may be considered with some property associated with points x on a line and the integral is carried out for the differential with two or more independent variables (x, y) as

$$\int_{x_0, y_0}^{x_1, y_1} F(x)\,dx = \int_{x_0, y_0}^{x_1, y_1} (A(x,y)\,dx + B(x,y)\,dy)$$

(10.62)

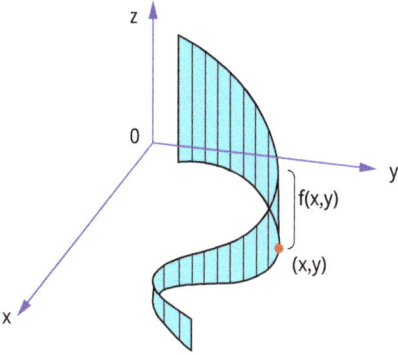

Figure 10.3: Line integral: Area of curved surface.

But this integral is not well defined, and if it is integrated with a pair of variables (x,y) in the xy plane, then many different paths between (x_0, y_0) and (x_1, y_1) can be used to join the two values in the plane. One must define the path along which the integration needs to be carried out. Then the integration may be represented as

$$\int_C F(x)dx = \int_C (A(x,y)dx + B(x,y)dy) \tag{10.63}$$

where C represents the curve joining the two points, and the integral is called a line integral:

$$\int_C F(x)dx = \int_{x_0}^{x_1} A(x,y)dx + \int_{y_0}^{y_1} B(x,y)dy \tag{10.64}$$

Example 7: Find the value of the integral

$$\int_C F(x)dx = \int_{x_0}^{x_1} (3x + 5y)dx + \int_{y_0}^{y_1} (2x + 7y)dy \tag{10.65}$$

where C is the straight line from $(0, 3)$ to $(2, 5)$ as follows:

$$y = 5x + 8 \tag{10.66}$$

Solution: To integrate the above line integral, the value of y is replaced by $5x + 8$ in the first part of integral, and in the second part of integral, x is replaced by $(y - 8)/5$ and then the integration is carried out over the limits as

$$\int_C F(x)dx = \int_0^2 (3x + 5(5x + 8))dx + \int_3^5 (2/5(y - 8) + 7y)dy \tag{10.67}$$

$$\int_C F(x)dx = \int_0^2 (28x + 40)dx + \int_3^5 ((37/5)y - 16/5)dy \tag{10.68}$$

$$\int_C F(x)dx = (14x^2 + 40x)\Big|_0^2 + ((37/10)y^2 + 16/5y)\Big|_3^5 \tag{10.69}$$

$$= 194.75$$

10.5.1 Multiple integral

When the integrand is to be integrated over two or more variables, it is called multiple integrals. It may be considered as an integral in higher dimensional space. In multiple integration, all but one variable is treated as constant and the integrals are integrated iteratively. A double integral may be defined as the limit of sums. A double integral may be considered as two single integrals which are integrated over different variables:

$$I = \iint f(x,y)dydx \tag{10.70}$$

First, the integral is integrated as $\int f(x,y)dy$ over the y interval (taking x as constant), and then this function is integrated over x interval (taking y as constant).

Example 8: Evaluate the double integral

$$\int_0^3 \int_0^2 (4 - y^2)dydx \tag{10.71}$$

Solution:

$$I = \int_0^3 \int_0^2 (4 - y^2)dydx$$

$$I = \int_0^3 \left(4y - \frac{y^3}{3}\right)\Big|_0^2 dx \tag{10.72}$$

$$I = \int_0^3 \left(8 - \frac{8}{3}\right)dx \tag{10.73}$$

$$I = \int_0^3 \frac{16}{3}dx \tag{10.74}$$

$$I = \left(\frac{16x}{3}\right)\Big|_0^3$$ (10.75)

$$I = 16$$

The order of integral can also be reversed but one has to change the order of integral limits as follows:

$$I = \int_0^2\int_0^3 (4 - y^2)\,dx\,dy$$ (10.76)

$$I = \int_0^2 (4x - xy^2)\Big|_0^3 dy$$ (10.77)

$$I = \int_0^2 (12 - 3y^2)\,dy$$ (10.78)

$$I = \left(12y - \frac{3y^3}{3}\right)\Big|_0^2$$ (10.79)

$$I = 16$$

10.5.2 Double integrals as volumes

When $f(x,y)$ is a function over rectangular region in xy plane, then its double integral can be interpreted as the volume of 3D region over xy plane bound by the R region and z plane (Fig. 10.4):

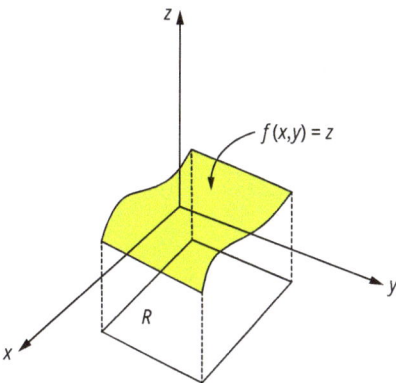

Figure 10.4: Double integrals as volume over xy plane bound by R region and z plane.

$$V = \iint_R f(x,y)\,dA$$ (10.80)

Fubini's theorem for calculating double integrals
According to this theorem, the double integral of any continuous function over a rectangle R can be calculated as an iterated integral w.r.t one variable at a time in any order. In this case, this double integral implies volume. One may calculate the volume by integrating in either order of x and y.

Example 9: Integrate the following

$$\iint_D x^2 y \, dA \tag{10.81}$$

where D is the rectangle defined by $0 \leq x \leq 1$ and $0 \leq y \leq 2$.
Solution: The above integral can be written as

$$I = \int_0^2 \int_0^1 x^2 y \, dx \, dy \tag{10.82}$$

$$I = \int_0^2 \left(\frac{x^3}{3} y \Big|_0^1 \right) dy \tag{10.83}$$

$$I = \int_0^2 \frac{y}{3} \, dy \tag{10.84}$$

$$I = \left(\frac{y^2}{6} \Big|_0^2 \right) \tag{10.85}$$

$$I = \frac{2}{3}$$

The integral can also be integrated in reverse direction, that is, first w.r.t y and then x

$$I = \int_0^1 \int_0^2 x^2 y \, dx \, dy \tag{10.86}$$

$$I = \int_0^1 \left(\frac{y^2}{2} x^2 \Big|_0^2 \right) dx \tag{10.87}$$

$$I = \int_0^1 2x^2 \, dx \tag{10.88}$$

$$I = \left(\frac{2x^3}{3} \Big|_0^1 \right) \tag{10.89}$$

$$I = \frac{2}{3}$$

10.5.3 Line integral of exact and inexact differentials

The line integrals of exact differentials depend on the initial (x_0, y_0) and final values (x_1, y_1) and not on the path of the function (i.e. the curve joining these points), that is, they are path independent:

$$\int_C F(x)dx = \int_C \left[\left(\frac{\partial F}{\partial x} \right) dx + \left(\frac{\partial F}{\partial y} \right) dy \right] \tag{10.90}$$

$$\int_C F(x)dx = F(x_1, y_1) - F(x_0, y_0) \tag{10.91}$$

If F is an inexact differential, then the line integral will depend not only on both the initial and final points but also on the path of integration.

Example 10: Evaluate if the following differential over the interval (0,0) to (1,1) is exact or inexact for the path $x = y$ and $y = x^2$:

$$\int F(x)dx = \int 2xydx + x^2 dy \tag{10.92}$$

Solution:

$$\Delta F = \int_0^1 2xydx + \int_0^1 x^2 dy \tag{10.93}$$

For the function $x = y$

$$\Delta F = \int_0^1 2x^2 dx + \int_0^1 y^2 dy \tag{10.94}$$

$$\Delta F = \frac{2x^3}{3} \Big|_0^1 + \frac{y^3}{3} \Big|_0^1 \tag{10.95}$$

$$\Delta F = \frac{2}{3} + \frac{1}{3} = 1$$

$$\Delta F = \int_0^1 2xydx + \int_0^1 x^2 dy \tag{10.93}$$

For the function $y = x^2$

$$\Delta F = \int_0^1 2x^3 dx + \int_0^1 ydy \tag{10.96}$$

$$\Delta F = \frac{2x^4}{4} \Big|_0^1 + \frac{y^2}{2} \Big|_0^1 \tag{10.97}$$

$$\Delta F = \frac{1}{2} + \frac{1}{2} = 1$$

Here the integrals of the differentials are equal for two different paths, hence the differentials are exact.

Example 11: Evaluate if the following differential over the interval (0,0) to (1,1) is exact or inexact for the path $x = y$ and $y = x^2$:

$$\int F(x)\,dx = \int xy\,dx + xy\,dy \tag{10.98}$$

Solution:

$$\Delta F = \int_0^1 xy\,dx + \int_0^1 xy\,dy$$

For the function $x = y$

$$\Delta F = \int_0^1 x^2\,dx + \int_0^1 y^2\,dy \tag{10.99}$$

$$\Delta F = \frac{x^3}{3}\Big|_0^1 + \frac{y^3}{3}\Big|_0^1 \tag{10.100}$$

$$\Delta F = \frac{1}{3} + \frac{1}{3} = \frac{2}{3}$$

$$\Delta F = \int_0^1 xy\,dx + \int_0^1 xy\,dy \tag{10.98}$$

For the function $y = x^2$

$$\Delta F = \int_0^1 x^3\,dx + \int_0^1 y^{3/2}\,dy \tag{10.101}$$

$$\Delta F = \frac{x^4}{4}\Big|_0^1 + \frac{2y^{5/2}}{5}\Big|_0^1 \tag{10.102}$$

$$\Delta F = \frac{1}{4} + \frac{2}{5} = \frac{13}{20}$$

Here the integrals of the differentials are different for two different paths, hence the differentials are inexact.

10.5.4 Line integrals in thermodynamics

In thermodynamics, a reversible process or a cyclic process is the one in which the starting and ending points are the same as the system returns to its original state. Hence, a cyclic process can be represented in the line integral as follows:

$$\oint dz = 0 \qquad\qquad (10.103)$$

Here z could be any thermodynamic parameter over the cyclic process and the line integral of the cyclic differential must be zero if it is exact. If dz is inexact, the line integral would not be equal to zero.

Example 12: Evaluate the cyclic integral of work done for the reversible process for 1 mol of an ideal gas where the gas is expanded from 0.01 to 0.02 m³ at constant temperature of 298 K at which the pressure is 3,167 Pa. The system is further subjected to the rise in temperature to 373 K at constant volume of 0.02 m³. The system is then compressed to a volume of 0.01 m³ at a temperature of 373 K at which the pressure is 101,325 Pa. The system is then cooled from 373 to 298 K at a constant volume of 0.01 m³.
Solution: The following cyclic process can be understood using Figure 10.5:

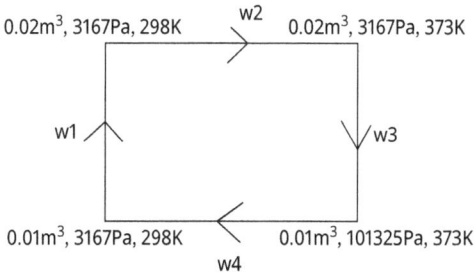

Figure 10.5: Cyclic process.

Step 1:
$$\int dw = -\int p\,dV$$
$$w = -p\Delta V$$
$$w_1 = -3,167 \times (0.02 - 0.01)\,\text{J}$$
$$w_1 = -31.67\,\text{J}$$

Step 2:
$$\int dw = -\int p\,dV$$
$$w = -p\Delta V$$
$$w_2 = 0 \text{ (as change in volume is zero)}$$

Step 3:
$$\int dw = -\int p\,dV$$

$$w = -p\Delta V$$

$$w_3 = -101{,}325 \times (0.01 - 0.02)\ \text{J}$$

$$w_3 = 1{,}013.25\ \text{J}$$

Step 4:

$$\int dw = -\int pdV$$

$$w = -p\Delta V$$

$$w_4 = 0\ \text{(as change in volume is zero)}$$

Hence the total work done is $w = w_1 + w_2 + w_3 + w_4$

$$w = 981.58\ \text{J}$$

The cyclic integral is not zero, which means that the work is an inexact differential which we have shown earlier as well.

10.6 Problems for practice

1. For an ideal gas undergoing isothermal process, the change in internal energy $dU = 0$. Evaluate w and q for the isothermal reversible expansion of 1.000 mol of a monatomic ideal gas from a volume of 15.50 L to a volume of 24.40 L at a constant temperature of 298 K.

2. Calculate the integral $I = \int_1^2 \int_y^{y^2} dxdy$

3. Evaluate the volume enclosed above the rectangle R as

$$V = \iint_R f(x,y)\,dA$$

$f(x,y) = 1 - 6x^2 y$ and $R: 0 \le x \le 2,\ -1 \le y \le 1$

Appendix A

Conversion factors between metric, US customary and British Imperial unit system.

Table A.1: Units of volume.

US or imperial		Metric
1 cu inch (in^3)		16.387 cm^3
1 cu foot (ft^3)		0.02832 m^3
1 fluid ounce	1.0408 UK fl oz	29.574 mL
1 pint (16 fl oz)	0.8327 UK pt	0.4732 L
1 gallon (231 in^3)	0.8327 UK gal	3.7854 L

Table A.2: Units of length.

US or imperial		Metric
1 inch (in)	12 in	2.54 cm
1 foot (ft)	3 ft	0.3048 m
1 yard (yd)	1,760 yd	0.9144 m
1 mile	12 in	1.6093 km
1 int. nautical mile	2,025.4 yd	1.853 km

Table A.3: Units of mass.

US or imperial		Metric
1 ounce (oz)	437.5 grain	28.35 g
1 pound (lb)	16 oz	0.4536 kg
1 stone	14 lb	6.3503 kg
1 hundredweight (cwt)	112 lb	50.802 kg
1 short ton (US)		0.9072 t
1 long ton (UK)		1.0160 t

https://doi.org/10.1515/9783111334448-011

Table A.4: Units of area.

US or imperial		Metric
1 sq inch (in^2)		6.4516 cm^2
1 sq foot (ft^2)	144 in^2	0.0929 m^2
1 sq yd (yd^2)	9 ft^2	0.8361 m^2
1 acre	4,840 yd^2	4,046.9 m^2
1 sq mile (mile^2)	640 acres	2.59 km^2

Table A.5: Prefixes for multiple and fractions of units.

Multiple	Prefix	Abbreviation	Fractions	Prefix	Abbreviation
10^1	Deca	da	10^{-1}	Deci	d
10^2	Hecto	h	10^{-2}	Centi	c
10^3	Kilo	k	10^{-3}	Milli	m
10^6	Mega	M	10^{-6}	Micro	μ
10^9	Giga	G	10^{-9}	Nano	n
10^{12}	Tera	T	10^{-12}	Pico	p
10^{15}	Peta	P	10^{-15}	Femto	f
10^{18}	Exa	E	10^{-18}	Atto	a
10^{21}	Zetta	Z	10^{-21}	Zepto	z
10^{24}	Yotta	Y	10^{-24}	Yocto	y

Table A.6: Volume unit conversion.

1 mL	0.001 L
1 cL	0.01 L
1 dL	0.1 L
1 in^3	1.639×10^{-2} L
1 gal	3.787 L
1 ft^3	28.316 L

Table A.7: Length unit conversion in metric system.

1 mm	0.001 m
1 cm	0.01 m
1 dm	0.1 m
1 dam	10 m
1 km	1,000 m
1 Amstrong	10^{-10} m
1 fermi	10^{-15} m
1 light year	0.946×10^{-16} m
1 mile	1.60937 m

Table A.8: Mass unit conversion in metric.

1 mg	0.001 g
1 cg	0.01 g
1 dg	0.1 g
1 dag	10 g
1 hg	100 g
1 kg	1,000 g
1 stone	6,350.29 g
1 lb	453.592 g
1 oz	28.3495 g

Table A.9: Units of time.

1 min	**60 s**
1 h	60 min/3,600 s
1 day	24 h
1 week	7 days
1 year	365 days

Table A.10: Energy unit conversion.

1 British thermal unit (BTU)	1,055 J
1 erg	10^{-7} J
1 foot print	1.356 J
1 calorie (cal)	4.186 J
1 kilowatt hour (kWh)	3.6×10^6 J
1 electron volt (eV)	1.602×10^{-19} J
1 litre atmosphere	101.13 J

Table A.11: Area unit conversion.

1 in^2	6.54×10^{-4} m^2
1 ft^2	9.2903×10^{-2} m^2
1 acre	4.0465×10^3 m^2
1 hectare	10^4 m^2
1 mile^2	2.5888×10^6 m^2
1 barn	10^{-28} m^2

Appendix B: Answers to the problems for practice

Unit 1

1. (a) 5 (b) 2 (c) 3 (d) 4 (e) 6 (f) 3 (g) 4
2. (a) 6,510,000 (b) 0.07 (c) 92.4 (d) 0.00125 (e) 0.0050 (f) 5.1
   (g) 30,000, 32,000, 32,400, 32,390, 32,391 (h) 80, 83.8, 83.8105 (i) 0.76
   (j) 1 (k) 900 (l) 8 (m) 40 (n) 0.02478 (o) 8500 (p) 0.00032
3. (a) 81.9 (b) 0.07194 (c) 3 (d) 2.073 (e) 2.4 (f) 35.6 (g) 1.0 × 105
   (h) 0.320 × 1011 (i) 170 (j) 0.09177

Unit 2

1. (a) 23.06 0.03 (b) 0.0137 0.0024 (c) 11.27 0.04 (d) (9 3) × 10–33
2. (a) 3.1417 is more precise while π is more accurate
   (b) 2.7182820135423 is more precise while 2.718281828 is more accurate

Unit 3

1. (a) $\log_2 60$ (b) $\frac{\log \pi}{\log 2}$ (c) $-5 \log_3 5$
2. $x = 1, y = 3$
3. $(x - 2)^2 + (y - 3)^2 = 25$
4. $f(V) = V^3 - (nb + \frac{nRT}{p})V^2 + \frac{n^2 a}{p} V - \frac{n^3 ab}{p} = 0$
5. $f(\Lambda_m) = c = \frac{1}{b^2} (\Lambda_m^\infty - \Lambda_m)^2 = 0$

Unit 4

1. 7.007, 0.191, 0.257, 0.663
2. $Q = 0.07 < Q_c$; hence, it can be included
3. $5.6 \pm 0.12\%$
4. $t = 0.424 < t_c$; hence, statistically correct
5. On the basis of F-test, $F = 0.375$ (compare with F_{tab} for $v_1 = 5$ and $v_2 = 7$)

Unit 5

1. 6.5 kJ mol^{-1}
2. $a = 26.694, b = 0.0094, c = -10^{-6}$

https://doi.org/10.1515/9783111334448-012

3. 6.90×10^{-4}
4. 0.0112 min^{-1}
5. $y = x^2$
6. 0.0111

Unit 6

1. With 10 intervals, 37,575; correct answer is 37,500
2. 1,444.2 with 20 intervals; correct answer is 1,443.1
3. $11.163 \text{ kJ mol}^{-1}$
4. $6.754 \text{JK}^{-1} \text{mol}^{-1}$
5. $40.53 \text{ kJ mol}^{-1}$

Unit 8

1. $\frac{dy}{dx} = 14$ and $\frac{d^2y}{dx^2} = 10$
2. 400 m s^{-1}
3. 5.687
4. 3.208

Unit 9

1. (a) -0.701367 (b) 1.855
2. 5.317
3. (a) 1.4142156 (b) 1.5707963
4. $1.649218 \times 10^{-7} \text{ M}$
5. 1.4375
6. (a) 0.517 (b) 1.86 (c) 0.565
7. 1.222
8. 0.5 and 0.70136
9. 1.85558
10. (a) 1.24682 (b) 1.4973

Unit 10

1. $w = -q = -1124.38 \text{ J}$
2. $5/6$
3. 4

Index

https://doi.org/10.1515/9783111334448-013